Novel Physical Processing of Starch-Based Products

From the perspective of physical action and materials science, this book reviews the structural modification of starch and evaluates novel physical processing of starch-based products.

The contributors systematically explain the multi-scale structures of starches and explore various treatments of starch and starch-based materials, including hydrothermal treatment, high-pressure treatment, extrusion treatment, electric field treatment, microwave treatment, ultrasonic treatment, cold plasma treatment, and 3D printing. By thoroughly analyzing and summarizing the structural control of starch, they aim to obtain starch-based products with better physicochemical properties and texture. Furthermore, their industrial applications for future food supply are also demonstrated.

This book will benefit researchers and graduate students in the fields of starch, starch-based materials, and food and non-food processing.

Enbo Xu is an associate research fellow at the College of Biosystems Engineering and Food Science, Zhejiang University. He focuses on carbohydrate utilization, food structure and processing, grain nutrition and function, etc. He has published about 90 papers in international journals and obtained 15 national invention patents.

Dandan Li is an associate professor at the College of Food Science and Technology, Nanjing Agricultural University. She focuses on starch modification, starch-based food production, nonthermal food processing technology, etc. In the past 5 years, she has published more than 70 papers.

Yang Tao is an associate professor at the College of Food Science and Technology, Nanjing Agricultural University. He focuses on physical processing of fruits and vegetables and food process simulation. In the past 5 years, he has published 50 papers in international journals.

Novel Physical Processing of Starch-Based Products

Strategy for Structure Control

Edited by
Enbo Xu, Dandan Li, and Yang Tao

CRC Press
Taylor & Francis Group
Boca Raton London New York

CRC Press is an imprint of the
Taylor & Francis Group, an **informa** business

This book is published with financial support from the National Key Research and Development Program of China (No. 2022YFF1101800), the Natural Science Foundation of China (No. 32272464 and No. 32102131), Young Elite Scientists Sponsorship Program by CAST (2022QNRC001), China Postdoctoral Science Foundation (2020M681631) and Jiangsu University Qinglan Project

Designed cover image: @Siyu Yao and Haohao Hu

First edition published 2025
by CRC Press
2385 NW Executive Center Drive, Suite 320, Boca Raton FL 33431

and by CRC Press
4 Park Square, Milton Park, Abingdon, Oxon, OX14 4RN

CRC Press is an imprint of Taylor & Francis Group, LLC

Library of Congress Cataloging-in-Publication Data
Names: Xu, Enbo, 1991- editor. | Li, Dandan, 1994- editor. | Tao, Yang, 1985- editor.
Title: Novel physical processing of starch-based products : strategy for
structure control / edited by Enbo Xu, Dandan Li, and Yang Tao
Description: Boca Raton, FL : CRC Press, 2025 | Includes bibliographical references
Identifiers: LCCN 2024004042 (print) | LCCN 2024004043 (ebook) |
ISBN 9781032797373 (hardback) | ISBN 9781032797403 (paperback) |
ISBN 9781003493594 (ebook)
Subjects: LCSH: Starch—Processing.
Classification: LCC TP248.S7 N68 2025 (print) | LCC TP248.S7 (ebook) |
DDC 664/.22—dc23/eng/20240326
LC record available at https://lccn.loc.gov/2024004042
LC ebook record available at https://lccn.loc.gov/2024004043

ISBN: 978-1-032-79737-3 (hbk)
ISBN: 978-1-032-79740-3 (pbk)
ISBN: 978-1-003-49359-4 (ebk)

DOI: 10.1201/9781003493594

Typeset in Minion
by codeMantra

Contents

Contributors

Ankutse Peter
Nanjing Agricultural University, Nanjing, Jiangsu Province, China

Dandan Li
Nanjing Agricultural University, Nanjing, Jiangsu Province, China

Enbo Xu
Zhejiang University, Hangzhou, Zhejiang Province, China

Haibo Pan
Zhejiang University, Hangzhou, Zhejiang Province, China

Hebin Xu
Nanjing Agricultural University, Nanjing, Jiangsu Province, China

Huifang Cao
Zhejiang University, Hangzhou, Zhejiang Province, China

Jie Li
Zhejiang University, Hangzhou, Zhejiang Province, China

Minxuan Chen
Zhejiang University, Hangzhou, Zhejiang Province, China

Na Yang
Jiangnan University, Wuxi, Jiangsu Province, China

Pei Wang
Nanjing Agricultural University, Nanjing, Jiangsu Province, China

Qingqing Zhu
Zhejiang University, Hangzhou, Zhejiang Province, China

Qisya Izanti Binti Mohammad Amirul Mursyid
University of Nottingham Malaysia Kuala Lumpur, Jalan Broga, Selangor, Malaysia

Shuohan Ma
Zhejiang University, Hangzhou, Zhejiang Province, China

Siyu Yao
Zhejiang University, Hangzhou, Zhejiang Province, China

Tian Ding
Zhejiang University, Hangzhou, Zhejiang Province, China

Wen Ma
Zhejiang University, Hangzhou, Zhejiang Province, China

Xiuqin Chen
Zhejiang University, Hangzhou, Zhejiang Province, China

Xuejiao Zhang
Nanjing Agricultural University, Nanjing, Jiangsu Province, China

Yanfeng Ding
Nanjing Agricultural University, Nanjing, Jiangsu Province, China

Yang Tao
Nanjing Agricultural University, Nanjing, Jiangsu Province, China

Yao Zhang
Jiangsu University of Science and Technology Zhenjiang, Zhenjiang, Jiangsu Province, China

Yongbin Han
Nanjing Agricultural University, Nanjing, Jiangsu Province, China

Zhengzong Wu
Qilu University of Technology, Jinan, Shandong Province, China

Acknowledgments

This work was supported jointly by the National Key Research and Development Program of China (No. 2022YFF1101800), the Natural Science Foundation of China (No. 32272464 and No. 32102131), Young Elite Scientists Sponsorship Program by CAST (2022QNRC001), China Postdoctoral Science Foundation (2020M681631), and Jiangsu University Qinglan Project.

CHAPTER 1

Multi-Scale Structure of Native and Processed Starches

Minxuan Chen, Enbo Xu, Qingqing Zhu, Zhengzong Wu, and Siyu Yao

1.1 INTRODUCTION

Starch, as energy storage polysaccharides in nature, is mainly found in seeds, roots and tubers of green plants. They also accumulate in stems, leaves, fruits and pollen (Pérez et al., 2010). Starch granules exhibit many shapes (polygonal, spherical, oval, elongated, lenticular and other irregular shapes) and sizes (in the range of ~0.1–200 μm, e.g., maize starch 5–20 μm, small wheat starch variants 2–3 μm, large wheat starch variants 22–26 μm, potato starch 15–75 μm, rice starch 3–8 μm, legumes starch 10–45 μm) in different starch types (Hoover, 2001; Pérez et al., 2010; Srichuwong et al., 2005; Tao et al., 2017; Tester et al., 2004). The physicochemical properties and applications of starches and starch-based materials depend on the molecular (amylose and amylopectin contents) and condensed structure characteristics, as well as the crystallinity (crystal size and area proportion), which vary with their botanic origins. Thus, the multi-scale structure of starch is an important factor during its processing such as hydrothermal treatment, extrusion, ultrasonic treatment, electromagnetism, plasma and other physical fields. As a granular form, natural starches

DOI: 10.1201/9781003493594-1

can be purposefully separated from proteins and lipids, and further green modified via physical and enzymatic processes to improve their original poor physicochemical and functional properties for food and non-food applications (Wei et al., 2018; Wu et al., 2017).

In this chapter, we introduce the latest researches on the multi-scale structure of natural starch macromolecules (or called supermolecules somtimes), especially the molecular/chain basis, lamellar structure and crystalline structure of starches. Then, the effects of physical action on the multi-scale structure of starch were simply summarized (detailed in the subsequent chapters). We also summarize the characterization methods of starch multi-scale structure changes, which is conducive to understanding of starch's internal fine structure and further promoting the development of natural starch polymer green processing.

1.1.1 Hierarchical Architecture of Native Starch

1.1.1.1 The Fine Structure of Amylose and Amylopectin

The hierarchical architecture of starch mainly includes granules (1–100 μm), growth rings (~0.1-0.5 μm), blocklets (20–50 nm), lamellar structures (9–10 nm), crystalline structures (several nm) and molecular structures (~0.1 nm). Amylose and amylopectin constitute the molecular structure (or called glucan chain structure) of starch as illustrated in Figure 1.1, and their fine structures and composition ratio determine the multi-scale structures and functionalities of starches (Tao et al., 2019). Amylose (20%–30% of normal starch granule) is essentially a linear polysaccharide that is made up of α-(1,4)-linked D-glucopyranosyl units, i.e., it has 500–5000 anhydro-glucose units per molecule, while amylopectin (70%–80%) is a highly branched polysaccharide which consists of α-(1,4)-linked D-glucopyranosyl backbones with α-(1,6)-linkage at their reducing ends (Bouvier and Campanella, 2014; Hoover, 2001; Kaur et al., 2012; Wolf, 2010). Amylopectin clusters form the thin crystalline layer of native starch, while the branch points of amylopectin, along with some sections of the disordered amylopectin chains, were primarily situated in amorphous lamellae. Generally, amylose is largely present in the amorphous lamellae, while only a small portion of amylose is involved in forming double helices with short side chains of amylopectin in the crystalline lamellae (Apriyanto et al., 2022). Amylose content greatly impacts the processibility of starch as well as the mechanical, thermal and rheological properties of starch-based products (Dean et al., 2007; Gilfillan et al., 2016; Liu et al., 2006, 2009; Stagner et al., 2011; Xie et al., 2013). In general, the

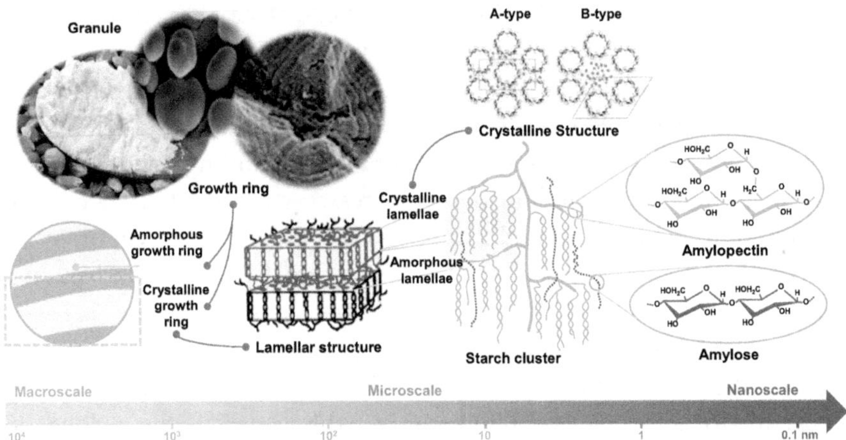

FIGURE 1.1 The multi-scale structure of starch.

higher the relative crystallinity and amylose content of starch, the stronger the resistance to physical processing (Dhital et al., 2011; Tan et al., 2015; Zhang et al., 2014), as well as to digestion.

1.1.1.2 The Multi-Scale Structure of Starch Granules

Amylose and amylopectin chains constitute the lamellar structure of starch granules. The semi-lamellar layer consists of a crystalline lamella (5–6 nm) and an amorphous lamella (3–4 nm). The semi-crystalline growth ring formed by alternating stacking of the two kinds of lamellae, together with the thinner and more width-uniform amorphous growth ring, constitutes the characteristic striped structure of starch granules, which can be observed under the optical microscope (Bouvier and Campanella, 2014; Pérez et al., 2010).

A three-dimensional multi-cylinder model based on French (1972) and Robin et al. (1974) has been used to describe the semi-crystalline structure formed by amylopectin cluster chains. The crystalline lamella is composed of ordered side chains of amylopectin with a double helix structure. The clusters of amylopectin helices are arranged into either A-, B- or C-type crystalline polymorphs, which can be detected by wide-angle X-ray diffraction (XRD) or solid-state ^{13}C nuclear magnetic resonance (NMR) spectroscopy. The V-type structure was recently defined and may be observed by processing, with the rotation of amylose chain forming a left spiral cavity structure. It is reported that the hydrophobic guests such as fat and phenolic substances could enter the spiral cavity of starch to form a V-type complex (Ma et al., 2022).

The A-type crystals, occurring mainly in cereal starches, have a monoclinic lattice pattern, more short amylopectin (DP<20) and are more densely packed, whereas B-type crystals with a hexagonal lattice pattern consist of longer amylopectin (DP>22) and have more associated water molecules, occurring predominantly in tubers (including potatoes), roots and high-amylose starches (or many processed starches). C-type crystals are the combination of A- and B-type crystals (the relative location of the two crystals differs with botanical sources), which are often in legumes, roots and some fruit and stem starches (Liu et al., 2022).

The amorphous lamellae of starch mainly consist of amylose, non-organized amylopectin chains (without helical configurations) and long linear amylopectin chains that interconnect the clusters, with amylopectin branch points. Compared with crystalline lamellae, amorphous lamellae have a lower bulk density of molecule chains, which makes them more sensitive to physicochemical and enzymatic treatments. In addition, the distribution of amylopectin branch points affects the processing sensitivity of the crystalline structure (Chi et al., 2021). A-type starch disperses its branch points in both amorphous and crystalline lamellae, while B-type starch clusters disperse in almost all branch points in amorphous lamellae.

The "blocklet" has been clearly defined as a structural element after being discovered these years (Gallant et al., 1997; Tang et al., 2006). The size of blocklets in the semi-crystalline growth rings ranges from 50 to 500 nm, with fewer structural imperfections than the smaller blocklets (20 nm) that make up the amorphous growth rings. Due to the difficulty of separation, the fine structure of the blocklets in natural starch has not been clearly described. Nonetheless, similar processing and digestive properties of blocklets to those of starch granules can be speculated due to the identical structural basis (i.e., periodic stacking of crystalline and amorphous lamellae).

1.1.2 Starch Structure Change and Its Identification Indexes

The functional properties of native and modified starch have received extensive attention in industry. The process-induced change and its mechanism of the formation of novel multi-scale structure have gradually been focused as for the achievement of functional diversification. During physical processing, the degree of interaction between amylose and amylopectin in both crystalline and amorphous regions changes slightly or greatly depending on the treating conditions. The common characterizations of the starch structure are concluded in Figure 1.2. In this section,

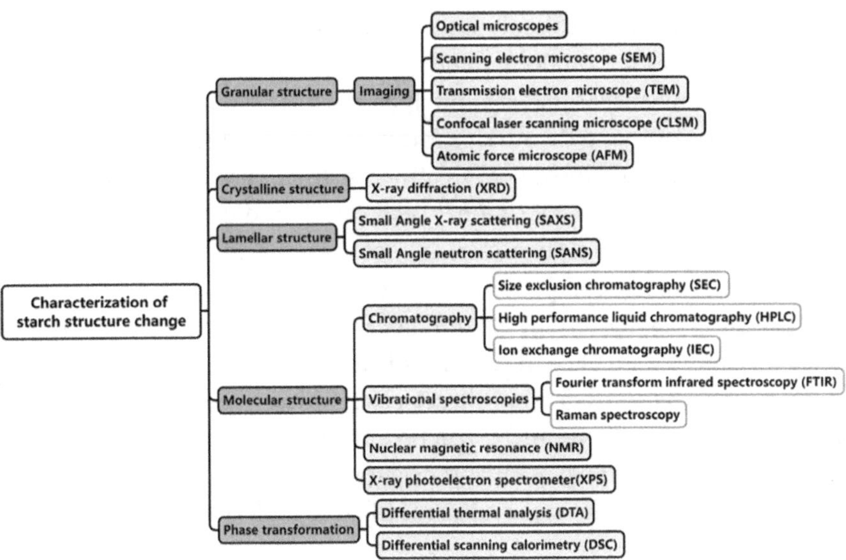

FIGURE 1.2 Characterization methods of starch structural change.

the analysis methods for starch granules morphology, crystal structure, lamellar structure and molecular structure are introduced.

1.1.2.1 Granule and Particle Morphology

The size of starch granules or processed particles (sometimes treated by powderization) is almost micrometer scale, so their microscopic morphologies can be observed by visualization techniques, including optical microscopy, scanning electron microscopy (SEM), transmission electron microscopy (TEM), atomic force microscopy (AFM) and confocal laser scanning microscopy (CLSM) (Chakraborty et al., 2020; Wu et al., 2023).

The optical microscopes used in the microstructure analysis of starch mainly include common biological optical microscope and polarized light microscope. The former can be used to observe the size and character of starches, the position of the umbilical point and the structure of growth rings, while the latter can be used to analyze the composition, position and change state of the polarization cross. The difference in density and refractive index between crystalline lamellae and amorphous lamellae leads to anisotropy, which generate polarization cross when polarized light passes through starches. The influence of different physicochemical factors on the lamellar structures can be inferred indirectly by analyzing the change of the polarization cross through the polarizing microscope.

Compared with optical microscopy, electron microscopy has the advantages of high resolution, long depth of field, large field of view, intuitive results, comprehensive analysis and dynamic observation of samples. SEM is suitable for studying the surface microstructure change of starches after physicochemical and biological treatments. In addition, SEM can also be used to observe the interaction between starch and other components to study the degradation characteristics of starch and starch-based complexes, the blending compatibility with other polymers, etc.

TEM has a higher resolution and more stringent requirements for sample preparation than SEM. It requires ultrathin sections or dilute solutions of starch samples rather than direct observation. TEM can provide the microstructure and sub-microstructure information of starches, including the structure arrangement and orientation of amorphous and crystalline lamellae, molecular size and shape of large molecular weight polymer, and atomic structures.

In recent years, CLSM (combining the advantages of optical and fluorescence microscopies) has been used to determine the microstructure and properties of starch granules or particles, as well as the structure of modified starch and starch-based materials. CLSM testing involves staining and slicing the sample, followed by multifaceted scanning to obtain an image of the inside of the sample. According to the fluorescence reaction between the dye and the end of starch reductive group, the reduced end of amylose was higher than that of amylopectin under the same molecular weight.

AFM provides surface structure and roughness information with nanoscale resolution by detecting the microscopic forces between the sample surface and the force-sensitive element. The atomic resolution makes AFM available for obtaining two-dimensional images of the surface and internal nanoscale structure of starches, as well as imaging starch molecular chains. When combined with microscopic analysis methods such as CLSM, the three-dimensional structure of samples can be analyzed with high precision (Low et al., 2020).

1.1.2.2 Crystal Structure

The structural order at starch semi-crystalline level is defined as long-range order, and its parameters and changes can be revealed by XRD technology, which can be divided into wide-angle X-ray scattering (WXRS) and small-angle X-ray scattering (SAXS) (Bouwman, 2021).

XRD produces continuous diffraction curves through periodically spaced atoms in starch crystals and has the advantages of simple sample preparation and non-destructive determination. The A-, B-, C- and V-type polymorphs of starch have respective characteristic diffraction peaks in the XRD pattern. The evolution of crystal structure, gelatinization, melting and phase transformation during physical processing, as well as the interaction of starch matrix composite system, can be interpreted by the parameters of the main diffraction peaks, including diffraction angle, peak width, diffraction intensity, relative diffraction intensity and interplanar crystal spacing (Chevigny et al., 2018).

Crystallinity, generally expressed by percentage, is an important index provided by the XRD pattern, which explicitly reflects the size of the degree of crystallization of the measured substance. So far, some X-ray diffractometers can give crystallinity information and determination simultaneously, while in more cases, it is necessary to artificially analyze and calculate according to the XRD pattern.

1.1.2.3 Lamellar Structure

Due to the presence of different electron cloud densities (ρ) between alternant amorphous and semi-crystalline layers, SAXS can be applied to characterize the lamellar structure properties of starches. The observation of lamellae suggests a relationship between the lamellar structure and the swelling of the blocklets (Salimi et al., 2023). The $I(q)$ of SAXS reflects the relationship between the relative peak intensity (I) and the scattering vector (q), and the original long period of native starch growth ring arrangement can be calculated as d_{Bragg}. On the other hand, according to the Lorentz correction method, the relationship between $\theta^2 I(\theta)$ and θ can be obtained, and then the long period of the newly formed crystalline lamellae can be calculated as d_{Lorentz}. The calculation formulas involved are as follows:

$$d = 2\pi / q$$

$$q = 4\pi \sin\vartheta / \lambda$$

2θ is the scattering angle and λ is the scattering wavelength (Cameron et al., 1992, 1993).

1.1.2.4 Molecular Structure

Chain length distribution (CLD) represents the quantity of the weight distribution of glucose units in starch chains and is expressed in different degree of polymerization (DP) values, which is a fundamental determinant of structure and strongly affects properties of starch-based products (Gebre et al., 2024). After the isoamylase pretreatment yielding starch branches, CLDs are usually measured by size exclusion chromatography or in combination with multi-angle laser scattering (MALS), fluorophore-assisted capillary electrophoresis, high-performance anion-exchange chromatography, and asymmetrical flow field-flow fractionation coupled with MALS (AF4-MALS) (Chi et al., 2021). The CLD of amylopectin side chains can be classified into four parts as follows: A (DP: 6–12), Bl (DP: 13–24), B2 (DP: 25–36), B3 (DP>36) and C chains (Bertoft et al., 2008), as well as some ultrashort chains (DP<6). The A chains are the outermost chains with no branching points, while the single C-chain per molecule is thought to form the backbone of the molecule, containing the sole reducing terminal residue (Chi et al., 2021). The weighted CLD pattern of amylose molecules in normal starches has been reported to be composed of three groups of peaks, denoted by AM1 consisting of shorter chains (DP 100–300), AM2 with intermediate-length chains (DP 300–1600) and long amylose chains (AM3, DP 1600–40,000) (Liu et al., 2022).

Starch chains randomly arranged in the amorphous region, but there is a small range of helical regular cross-linking between neighboring chain molecules, which is defined as the short-range ordering of starch (Chi et al., 2021). Fourier transform infrared spectroscopy (FTIR) and Raman spectrum, based on the determination of vibrational energy levels related to starch chemical bonds, are the main tools for detecting short-range and medium-range interactions in starch structures. In the FTIR spectrum, the absorption peak at 1047 cm^{-1} corresponds to the crystalline structure of starch aggregates, and the absorption peak at 1022 cm^{-1} corresponds to the irregular aggregation of starch macromolecules. The ratio of these two absorbance values ($R_{1047/1022}$, degree of order, DO) and the full width at half maximum (FWHM) at 480 cm^{-1} of Raman spectrum have been widely used to characterize the short-range molecular order of the double helix in starch (Ma et al., 2022; Mutungi et al., 2012).

NMR has the ability to simultaneously study the chemical structure and kinetic properties of starch chains in a non-destructive and non-invasive manner. At present, NMR techniques, including variable temperature NMR, solid-state NMR and time domain NMR, have been widely used to study the structural changes of starch at different levels, and to a large extent solve the shortcomings of low sensitivity in the complex system of this technology (Wu et al., 2023). XPS is a useful tool for exploring surface chemical changes at depths of 6–8 nm, allowing qualitative, quantitative and chemical state analyses of all elements except hydrogen. NMR and XPS can be used to determine the breaking and formation of chemical bonds in starch chains and reveal the nanostructure composition (Yao et al., 2023). In ^{13}C NMR spectra, C1 (100–110 ppm) is related to the type and range of crystallinity, while C2, C3, C5 (71–74 ppm) and C6 (58–63 ppm) can provide double- and single-helix content information, respectively (Yin et al., 2023). The ratio of O and C content (O/C) determined by XPS can be used to infer hydroxyl exposure. The changes in the intensity of C–O, C–O–H and other chemical bond peaks in XPS spectra can be used to explain the process of intermolecular interaction and molecular miscibility (Liang et al., 2020).

1.1.3 General Physical Modification of Starch

According to the treatment temperature, the physical modification methods of starch can be divided into thermal and non-thermal processes. The former, also known as hydrothermal treatment (Chapter 2), involves treating starch at various temperature–moisture combinations, with common types including atmospheric pressure gelatinizing, annealing that performed at excessive moisture content (>~40%) and temperature lower than starch gelatinization point, heat-moisture treatment that performed at low moisture content (usually 18%–27%) and high-temperature conditions and autoclaving that performed at an excessive moisture content and a high temperature. Another is extrusion (Chapter 4), which uses thermal energy combined with pressure, kneading, shearing, etc.

Non-thermal treatments of starch mainly include ultrahigh pressure (Chapter 3), electric field (Chapter 5), magnetic field (Chapter 6), ultrasound (Chapter 7) and cold plasma (Chapter 8), though sometimes they are also used in combination with heat treatment, or with the generation of thermal energy during processing. Non-thermal treatment has the extra advantages of reducing energy and water consumption and shortening processing times, and can to some extent change the function of starch without excessively affecting its sensory properties, which is a key necessity in food applications and therefore become a research hotspot in recent years. Overall, the above physical treatment methods of starch have not been described here as their specific research progress will be elaborated in the subsequent chapters of this book.

REFERENCE, BIBLIOGRAPHY OR WORKS CITED

Apriyanto, A., Compart, J., & Fettke, J. 2022. A review of starch, a unique biopolymer-structure, metabolism and in planta modifications. *Plant Science* 318: 111223.

Bertoft, E., Piyachomkwan, K., Chatakanonda, P., & Sriroth, K. 2008. Internal unit chain composition in amylopectins. *Carbohydrate Polymers* 74(3): 527–543.

Bouvier, J. M., & Campanella, O. H. 2014. *Extrusion processing technology: food and non-food biomaterials.* UK: Wiley Blackwell.

Bouwman, W. G. 2021. Spin-echo small-angle neutron scattering for multiscale structure analysis of food materials. *Food Structure* 30: 100235.

Cameron, R. E., & Donald, A. M. 1992. A small-angle X-ray scattering study of the annealing and gelatinization of starch. *Polymer* 33(12): 2628–2635.

Cameron, R. E., & Donald, A. M. 1993. A small-angle X-ray scattering study of the absorption of water into the starch granule. *Carbohydrate Research* 244(2): 225–236.

Chakraborty, I., Pallen, S., Shetty, Y., Roy, N., & Mazumder, N. 2020. Advanced microscopy techniques for revealing molecular structure of starch granules. *Biophysical Reviews* 12(1): 105–122.

Chevigny, C., Chaunier, L., Ferbus, R., Roblin, P., Rondeau-Mouro, C., & Lourdin, D. 2018. In-situ quantitative and multiscale structural study of starch-based biomaterials immersed in water. *Biomacromolecules* 19(3): 838–848.

Chi, C., Li, X., Huang, S., Chen, L., Zhang, Y., Li, L., & Miao, S. 2021. Basic principles in starch multi-scale structuration to mitigate digestibility: a review. *Trends in Food Science & Technology* 109: 154–168.

Dean, K., Yu, L., & Wu, D. Y. 2007. Preparation and characterization of melt-extruded thermoplastic starch/clay nanocomposites. *Composites Science and Technolog,* 67(3–4): 413–421.

Dhital, S., Shrestha, A. K., Flanagan, B. M., Hasjim, J., & Gidley, M. J. 2011. Cryo-milling of starch granules leads to differential effects on molecular size and conformation. *Carbohydrate Polymers* 84(3): 1133–1140.

French, D. 1972. Fine structure of starch and its relationship to the organization of starch granules. *Journal of the Japanese Society for Starch Science* 19: 8–12.

Gallant, D. J., Bouchet, B., & Baldwin, P. M. 1997. Microscopy of starch: evidence of a new level of granule organization. *Carbohydrate Polymers* 32(3): 177–191.

Gebre, B. A., Zhang, C., Li, Z., Sui, Z., & Corke, H. 2024. Impact of starch chain length distributions on physicochemical properties and digestibility of starches. *Food Chemistry* 435: 137641.

Gilfillan, W. N., Moghaddam, L., Bartley, J., & Doherty, W. O. S. 2016. Thermal extrusion of starch film with alcohol. *Journal of Food Engineering* 170: 92–99.

Hoover, R. 2001. Composition, molecular structure, and physicochemical properties of tuber and root starches: a review. *Carbohydrate Polymers* 45(3): 253–276.

Kaur, B., Ariffin, F., Bhat, R., & Karim, A. A. 2012. Progress in starch modification in the last decade. *Food Hydrocolloids* 26(2): 398–404.

Liang, Y., Zhu, H., Wang, L., He, H., & Wang, S. 2020. Biocompatible smart cellulose nanofibres for sustained drug release via pH and temperature dual-responsive mechanism. *Carbohydrate Polymers* 249: 116876.

Liu, H., Xie, F., Yu, L., Chen, L., & Li, L. 2009. Thermal processing of starch-based polymers. *Progress in Polymer Science* 34(12): 1348–1368.

Liu, H., Yu, L., Xie, F., & Chen, L. 2006. Gelatinization of cornstarch with different amylose/amylopectin content. *Carbohydrate Polymers* 65(3): 357–363.

Liu, X., Huang, S., Chao, C., Yu, J., Copeland, L., & Wang, S. 2022. Changes of starch during thermal processing of foods: current status and future directions. *Trends in Food Science & Technology* 119: 320–337.

Low, L. E., Siva, S. P., Ho, Y. K., Chan, E. S., & Tey, B. T. 2020. Recent advances of characterization techniques for the formation, physical properties and stability of Pickering emulsion. *Advances in Colloid and Interface Science* 277: 102117.

Ma, S., Zhu, Q., Yao, S., Niu, R., Liu, Y., Qin, Y., & Xu, E. 2022. Efficient retention and complexation of exogenous ferulic acid in starch: could controllable bio-extrusion be the answer? *Journal of Agricultural and Food Chemistry* 70(47): 14919–14930.

Mutungi, C., Passauer, L., Onyango, C., Jaros, D., & Rohm, H. 2012. Debranched cassava starch crystallinity determination by Raman spectroscopy: correlation of features in raman spectra with X-ray diffraction and 13C CP/MAS NMR spectroscopy. *Carbohydrate Polymers* 87(1): 598–606.

Pérez, S., & Bertoft, E. 2010. The molecular structures of starch components and their contribution to the architecture of starch granules: a comprehensive review. *Starch* 62(8): 389–420.

Robin, P. J., Mercier, C., Charbonnière, R., & Guilbot, A. 1974. Litnerized starches. Gel filtration and enzymatic studies of insoluble residues from prolonged acid treatment of potato starch. *Cereal Chemistry* 51: 389–406.

Salimi, M., Channab, B., El Idrissi, A., Zahouily, M., & Motamedi, E. 2023. A comprehensive review on starch: structure, modification, and applications in slow/controlled-release fertilizers in agriculture. *Carbohydrate Polymers* 322: 121326.

Srichuwong, S., Sunarti, T., Mishima, T., Isono, N., & Hisamatsu, M. 2005. Starches from different botanical sources I: contribution of amylopectin fine structure to thermal properties and enzyme digestibility. *Carbohydrate Polymers* 60(4): 529–538.

Stagner, J., Dias Alves, V., Narayan, R., & Beleia, A. 2011. Thermoplasticization of high amylose starch by chemical modification using reactive extrusion. *Journal of Polymers and the Environment* 19(3): 589–597.

Tan, X., Zhang, B., Chen, L., Li, X., Li, L., & Xie, F. 2015. Effect of planetary ball-milling on multi-scale structures and pasting properties of waxy and high-amylose cornstarches. *Innovative Food Science & Emerging Technologies* 30: 198–207.

Tang, H., Mitsunaga, T., & Kawamura, Y. 2006. Molecular arrangement in blocklets and starch granule architecture. *Carbohydrate Polymers* 63(4): 555–560.

Tao, H., Xiao, Y., Wu, F., & Xu, X. 2017. Optimization of additives and their combination to improve the quality of refrigerated dough. *LWT - Food Science and Technology* 89: 482–488.

Tao, K., Li, C., Yu, W., Gilbert, R. G., & Li, E. 2019. How amylose molecular fine structure of rice starch affects functional properties. *Carbohydrate Polymers* 204: 24–31.

Tester, R. F., Karkalas, J., & Qi, X. 2004. Starch-composition, fine structure and architecture. *Journal of Cereal Science* 39(2): 151–165.

Wei, B., Cai, C., Xu, B., Jin, Z., & Tian, Y. 2018. Disruption and molecule degradation of waxy maize starch granules during high pressure homogenization process. *Food Chemistry* 240: 165–173.

Wolf, B. 2010. Polysaccharide functionality through extrusion processing. *Current Opinion in Colloid & Interface Science* 15(1–2): 50–54.

Wu, C., Zhou, X., Tian, Y., Xu, X., & Jin, Z. 2017. Hydrolytic mechanism of α-maltotriohydrolase on waxy maize starch and retrogradation properties of the hydrolysates. *Food Hydrocolloids* 66: 136–143.

Wu, Z., Li, S., Chen, G., Wang, Y., & Li, H. 2023. Advanced in situ analytical techniques: opening the door for the structural analysis of starch and its food applications. *Trends in Food Science & Technology* 138: 57–73.

Xie, F., Pollet, E., Halley, P. J., & Avérous, L. 2013. Starch-based nano-biocomposites. *Progress in Polymer Science* 38(10–11): 1590–1628.

Yao, S., Ma, S., Zhu, Q., Qin, Y., Ngah, W., Zuo, Y., & Xu, E. 2023. Ultrasmooth and uniform starch nanosphere with new microstructure formation via microfluidization-nanoprecipitation control. *ACS Sustainable Chemistry & Engineering* 11(19): 7475–7488.

Yin, X., Hu, Z., Zheng, Y., Chai, Z., Kong, X., Chen, S., & Tian, J. 2023. Multi-scale structure characterization and in vivo digestion of parboiled rice. *Food Chemistry* 402: 134502.

Zhang, B., Zhao, Y., Li, X., Li, L., Xie, F., & Chen, L. 2014. Supramolecular structural changes of waxy and high-amylose cornstarches heated in abundant water. *Food Hydrocolloids* 35: 700–709.

CHAPTER 2

Hydrothermal Treatment of Starch and Starchy Materials

Yao Zhang and Pei Wang

2.1 INTRODUCTION

Starch has been broadly used in food, feed, chemical, pharmaceutical, and eco-friendly industries due to the merits of abundance, biodegradability, and nontoxicity, but those applications are limited by their native forms and botanical diversity (Maniglia et al., 2021). Modifications on starch structure imparted by physical treatments are green and efficient for alleviating the facing limitations. Hydrothermal treatments including heat-moisture treatment (HMT) and annealing (ANN) allow for changes in the packing arrangements of polymer molecules within starch granules and the overall structures of granular surfaces under controlled conditions of temperature, moisture, and duration (BeMiller & Huber, 2015). Therefore, they have been widely used to obtain starch with desired properties in the food and other non-food sectors over the past decades (Figure 2.1).

In HMT, starch is processed at a temperature higher than its gelatinization temperature (typically, 90°C–120°C) under a low water level (typically, 10%–30%, w/w), while in ANN, starch is treated between the glass transition and gelatinization temperatures in excess (typically, >60%) or intermediate (typically, 50%–60%) water contents (Jacobs & Delcour, 1998;

DOI: 10.1201/9781003493594-2

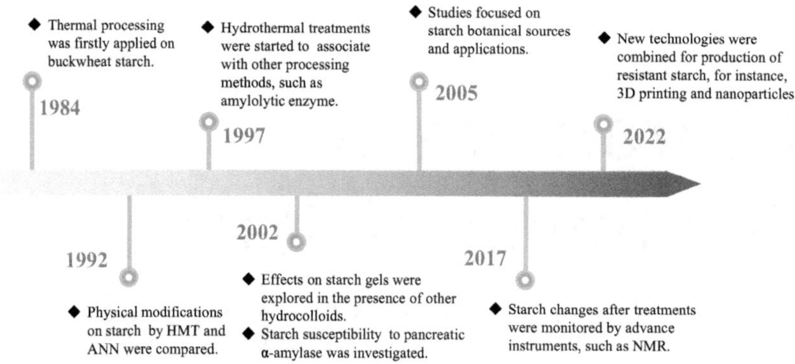

FIGURE 2.1 Timeline of the research progress of hydrothermal treatments on starch.

Tester & Debon, 2000; Zavareze & Dias, 2011). Hydrothermal treatment on starches could alter the mobility of amylose (AM) and amylopectin (AP) chains within the amorphous and crystalline domains without damaging starch granular structures, and modifying their physicochemical properties. Research geared toward hydrothermal processing on starch is important, since this process is green and safe, which is allowed by governments and preferred by consumers. Therefore, this chapter aims to summarize the principle of hydrothermal action on starch, the hydrothermal modifications on thermal stability and digestibility, and the potential applications of hydrothermal-treated starch (HTS) in industries such as food, film preparation, and so on.

2.2 PRINCIPLE OF HYDROTHERMAL ACTION

The underlying mechanism of hydrothermal action on starch has been proposed involving: (1) in crystalline regions, unstable structures being broken and highly ordered double helices being perfected, and (2) in amorphous areas, disruption of loose helical arrangements and new associations of AM–AM, AM–lipid, AM–AP or AP–AP being both induced (BeMiller & Huber, 2015; Iuga & Mironeasa, 2020) (Figure 2.2). These changes subsequently alter the morphology, crystalline structures, and molecular distributions of starch, which is discussed in Sections 2.2.1–2.2.3.

2.2.1 Granular Morphology and Size

Changes in the morphology and size depend on the starch origin and the applied hydrothermal conditions, especially the treated temperature

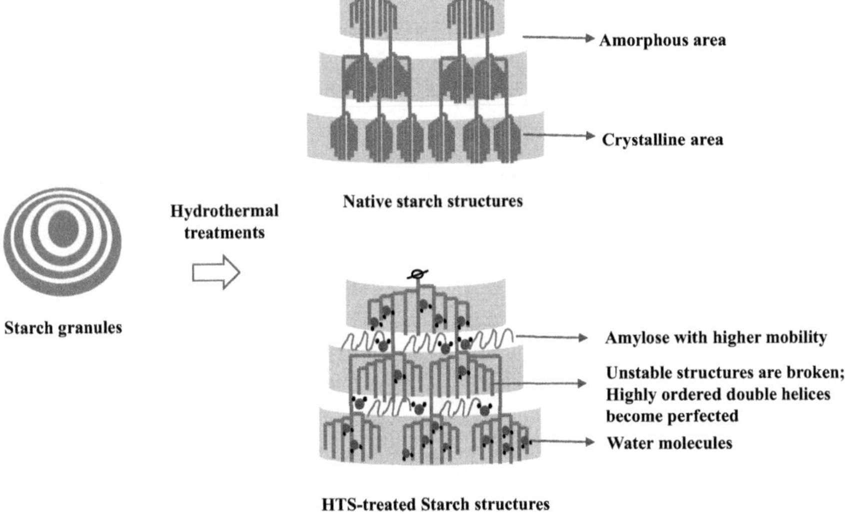

FIGURE 2.2 Chain mobility of native starch and hydrothermal-treated starch (HTS).

and moisture content (Zavareze & Dias, 2011), as illustrated in Figure 2.3. At low water contents, HMT did not change the granule shape and size of normal rice, tuber, and root starches (Arns et al., 2015; Gunaratne & Hoover, 2002). However, the imperfection of starch granules was often increased by HMT with a moisture content increase. For example, rice starch aggregated and their surface roughness increased at 25 g/100 g moisture content (Ruiz, Srikaeo, & Revilla, 2018). Banana and mango kernel starches by HMT also became inhomogeneous (Bharti, Singh, & Saxena, 2019; Cordeiro et al., 2018). For coix seed starch (Liu et al., 2016) and some bean starches (Ambigaipalan et al., 2014), fissures, pores, hollows, or granule fragments appeared. The particle diameter of HMT starches was increased or remained unchanged, which mainly resulted from the partial gelatinization and molecular restructuring (BeMiller & Huber, 2015). A similar appearance and dimension were observed before and after ANN, such as retrograded crystallites in debranched waxy rice starch (Ji et al., 2019) and new cocoyam starch (Lawal, 2005). However, the increase in particle sizes and appearance of surface irregularities were still observed following ANN, which was possibly due to the swelling of small granules under a moist environment (Lan et al., 2008; Molavi, Razavi, & Farhoosh, 2018).

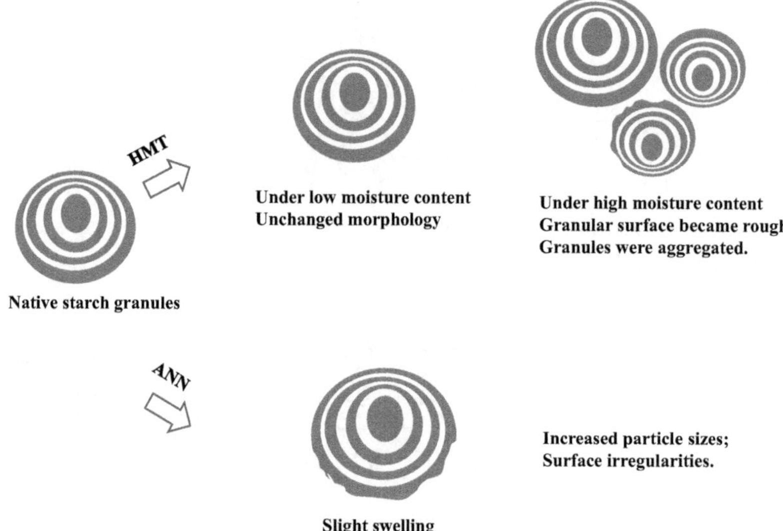

FIGURE 2.3 Effect of hydrothermal treatment on the granular morphology and size of starch.

2.2.2 Crystal Structure and Relative Crystallinity

The effect of HMT and ANN on the X-ray patterns and relative crystallinity of starches is depicted in Figure 2.4. Variations in X-ray diffractograms after HMT are determined by starch sources (Jayakody & Hoover, 2008). For normal wheat (Chen et al., 2015), rice (Wang et al., 2018a), and maize starches (Wang et al., 2016), their A-type crystal were not changed by HMT. A similar trend was claimed for pulse starches after HMT, such as mung bean, navy bean, and pea and lentil starches (Barua & Srivastav, 2017; Chung, Liu, & Hoover, 2010). A V-type packing arrangement was sometimes observed on cereal starches following HMT due to the formation of AM–lipid complexation (Chen et al., 2015; Kumar et al., 2023; Shi, Gao, & Liu, 2018). The B-type polymorph was partially or completely converted to the A-type by hydrothermal treatments. A coexistence of A- and B-crystals was shown when the B-type high-AM maize starch was subjected to HMT (Wang et al., 2016). For waxy and normal potato starches, the polymorphic transition of B- to A-types also occurred under the HMT temperature between 80°C and 130°C and the moisture of 27% for 16 hours, and the conversion was enhanced with increasing temperature (Varatharajan et al., 2011). A dehydration of the B-type unit cell was supposed to occur

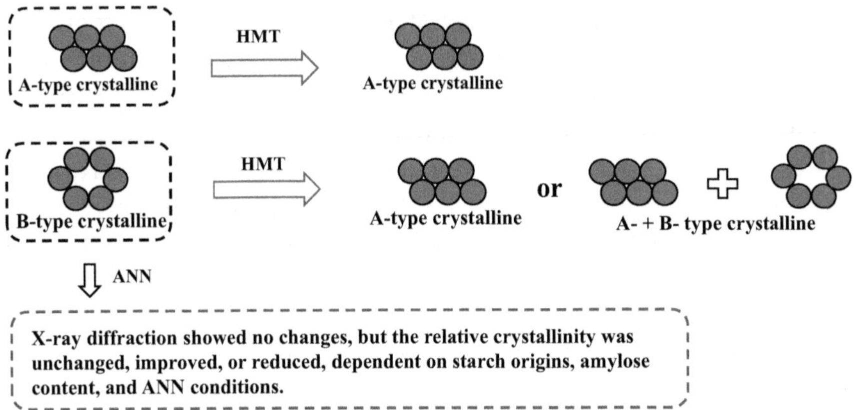

FIGURE 2.4 Effect of hydrothermal treatment on the semicrystalline structure of starch.

due to the applied thermal energy during HMT, and double helices could then move into the crystalline region, which resulted in a stronger hydrogen bonding and a more ordered structure (Gunaratne & Hoover, 2002; Pepe et al., 2016). Some authors also proposed a consistent mechanism for the disruption of hydrate water bridges and the crystalline arrangement (Hoover, 2010). Changes in the relative crystallinity largely depended on starch moisture levels. The relative crystallinity of wheat starch was increased by HMT at a low water level (15%–25%) because of higher molecular mobility and enhanced AM–AP interactions (Chen et al., 2015; Ma & Boye, 2018). A similar result of the relative crystallinity was also reported for tuber (yam and potato) starches after HMT (90°C–120°C, 16 hours) under a humid condition (20%–30% moisture) (Gunaratne & Hoover, 2002; Vermeylen, Goderis, & Delcour, 2006). However, an opposite behavior was shown on wheat and rice starches when more water was incorporated, which could be attributed to the gelatinization of starch granules and the disruption of crystalline areas (Chen et al., 2015; Silva et al., 2017).

ANN typically did not alter the diffraction pattern of starch, which differed from HMT. This different effect on starch structure and properties is directly related to the parameters applied in both modifications, e.g., moisture content, temperature, and exposure time. However, the relative crystallinity could be unchanged, improved, or reduced by ANN, which depends on starch sources, the AM content of starch granules, and ANN parameters. Samarakoon et al. (2020) reported that ANN affected the structural and physicochemical properties of starches from different botanical sources differently. The diffraction patterns of all starches showed no change, the relative crystallinity of barely, rice, and potato starches were not changed, but the relative crystallinity of corn starch was increased by 9%. AM content is important in determining the relative crystallinity of starch. Effects of ANN on corn starch with different AM contents were investigated by Rocha et al. (2012). Their results suggested that the diffraction patterns of starch with different AM contents was not changed after ANN and the relative crystallinity of normal corn starch showed no change, but the crystallinity for waxy starch was increased by 7%. ANN conditions also determined the variations in relative crystallinity. Li et al. (2020a) reported that the relative crystallinity of wheat and corn starches was decreased with increasing ANN temperature. Overall, the effect of hydrothermal modification is different due to the botanical source of starches. Hence, when HMT and/or ANN was applied to modify starch, this needs to be taken into consideration, since the modifications can promote specific outcomes for each starch type.

2.2.3 Thermal Degradation

During the high-temperature processing, degradation of starch polymers is favored. When HMT was conducted under 130°C, the amounts of ultra-short chains with a low degree of polymerization (DP$\leq$6) were increased. Meanwhile, more mobile short chains were available for molecular realignments and the formation of more perfect crystals (Vermeylen, Goderis, & Delcour, 2006). Likewise, an increased crystallinity of mung bean starch (Li, Ward, & Gao, 2011) and waxy rice starch (Zeng et al., 2015) after HMT was reported. Thermal degradation mainly occurred on double-helical segments from AP–AP and at or near α-1,6-branching linkages (Kim & Huber, 2013). ANN is normally useful in decreasing the molecular weights (M_w) of long starch chains (Shen et al., 2021).

2.3 CHANGES IN PHYSICOCHEMICAL PROPERTIES OF HTS

Thermal parameters, including starch swelling, pasting, and gelatinization, determine the rheological properties of starchy materials and final product qualities (Jia et al., 2023). Starch digestibility confers their functional properties. In this study, the positive effects of hydrothermal treatments on starch thermal and digestible stabilities are depicted in Figure 2.5.

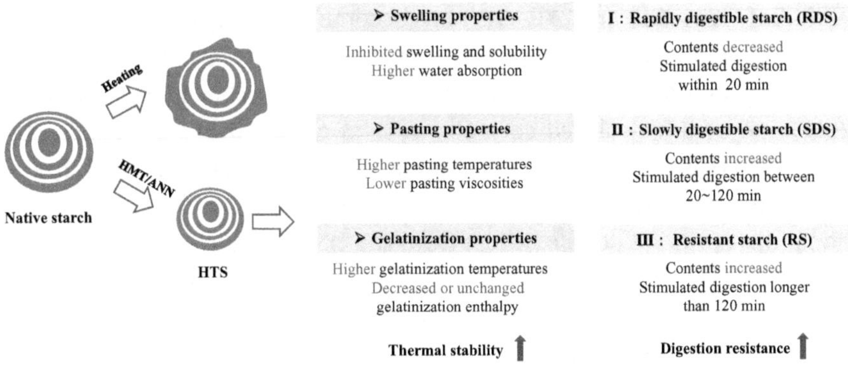

FIGURE 2.5 Changes in physiochemical properties of hydrothermal-treated starch (HTS).

2.3.1 Granular Swelling, AM Leaching, Solubility, and Water Absorption

The swelling behavior which is generally evaluated by swelling factor and swelling power is decided by the pasting temperature and granule components including AM and lipid contents and AP structures (BeMiller, 2011). The swelling factor measures the water within granules, and the swelling power is quantitatively related to both intra- and intergranular water. When the ANN was applied on normal and high-AM wheat starch, swelling factor decreased, being indicative of enhanced starch molecule associations and crystalline perfection (Lan et al., 2008). This was due to the promotion of ANN to additional interactions between AM–AM and AM–AP and formations of AM–lipid complexes (Tester, Debon, & Sommerville, 2000). After HMT, swelling power and AM leaching have most commonly been suppressed, for instance, white sorghum starch (Olayinka, Adebowale, & Olu-Owolabi, 2008), potato starch (Kim et al., 2013), cocoyam starch (Lawal, 2005), sword bean (Olu-owolabi, Afolabi, & Adebowale, 2011), and mango kernel starch (Bharti, Singh, & Saxena, 2019). Solubility of HMT-wheat and HMT-buckwheat was also found to be lower than the native counterpart, likewise accompanied by inhibited swelling (Li et al., 2019; Sindhu & Khatkar, 2023). These decreases are generally more obvious with a higher HMT temperature or moisture content. It was a result of the increased granular rigidity (Ambigaipalan et al., 2014; Bharti, Singh, & Saxena, 2019). AM could transform into helical structures or form complexes with lipid molecules in amorphous areas and interactions between

AM–AM, AM–AP, and AP–AP may become pronounced in crystalline regions (Chung, Liu, & Hoover, 2010; Suriya, Reddy, & Haripriya, 2019). The swelling power of ball-milling-modified wheat starch (Liu, Hong, & Zheng, 2017) and acid-hydrolyzed corn starch (Sun et al., 2015) was reduced by HMT, whereas their solubility increased. Certain damage to starch granules could decrease the granular stability, which enhanced the conformational disruption of double helices from the crystalline areas during HMT (Liu, Hong, & Zheng, 2017). Moreover, higher solubility could result from the stronger water-binding ability of amorphous areas or from the extension of intergranular spaces (Suriya, Reddy, & Haripriya, 2019). After HMT or ANN, sorghum and green banana starches presented high water absorption (Adebowale & Olu-Owolabi, 2011; Cordeiro et al., 2018). This indicated that granules after hydrothermal treatments had high water-binding capacity. Breakages of hydrogen bonds provided more mobile chains for hydration (Adebowale et al., 2011). For cassava starch, water absorption was increased after HMT, while it was decreased by ANN (Falade, Ibanga-Bamijoko, & Ayetigbo, 2019). This was probably due to differences in the exposure of hydrophilic sites.

2.3.2 Pasting Properties

Pasting temperature is the starting temperature at which the viscosity of starch suspension dramatically increases on heating, and peak, breakdown, and setback viscosities describe the swelling power, disintegration, and rapid retrogradation degrees, respectively (Chen et al., 2022). Higher pasting temperatures and lower viscosities were found for wheat starch (Chen et al., 2015), rice starch (Henning et al., 2019), cocoyam starch (Lawal, 2005), sweet potato starch (Liao et al., 2019), black and bayo bean starches (Chavez-Murillo, Orona-Padilla, & Rosa-Millan, 2019), and broad bean, chickpea, and lentil starches (Chávez-Murillo et al., 2018), and amaranth starch (Siwatch, Yadav, & Yadav, 2022) after HMT or ANN compared to their native counterparts. Hydrothermal treatments reinforce the interactions between starch and lipid molecules in the crystalline and amorphous regions (Iuga & Mironeasa, 2020). Therefore, these treated starches need more heat energy to initiate pasting and become more stable in the further pasting process. During paste cooling, less AM and short AP reassociate, which shows a retardation of the short-term retrogradation. The moisture level and treating duration are essential for the modification of pasting properties. More obvious changes were reported for HMT as compared with ANN (Molavi, Razavi, & Farhoosh, 2018). A similar trend was also observed for wheat, sorghum, buckwheat, and high β-glucan barley flours after HMT (Collar, 2017). However, the application of HMT increased viscosities of barley flour. Different pasting behaviors of various starches or flours are relevant to granular sizes, the AM content, and the presence of soluble and insoluble fibers (Chávez-Murillo et al., 2018; Loubes, González, & Tolaba, 2018).

2.3.3 Gelatinization Properties

The thermal properties are characterized by gelatinization parameters that are the onset (T_o), peak (T_o), and conclusion (T_c) temperatures and the gelatinization enthalpy (ΔH) (Ratnayake & Jackson, 2008). Amaranth starch registered increased T_o, T_o, and T_c values under HMT at moisture higher than 20% (Bet et al., 2018). During HMT modifications, increases in those characteristic temperatures were greater at higher moisture contents (Mustapha et al., 2019). Higher gelatinization temperatures were also required for maize starch (Wang et al., 2016), rice starch (Zavareze & Dias, 2011), and black bean, fava bean, chickpea, and lentil flours (Chávez-Murillo et al., 2018) after HMT or ANN. The improvement of granular thermal stability was proportional to the perfection of granular structures by the strengthening of double-helical structures between AM and AP and the limited mobility of amorphous areas due to AM–AM and AM–lipid interactions. HMT has been claimed to improve the thermal stability of starch than ANN. ΔH value has been shown to decrease or remain unchanged under HMT or ANN. Lower crystallinity, a disruption of the amorphous area, and partial gelatinization contributed to less heat energy required for melting crystalline clusters, which means a lower ΔH value (Puncha-arnon & Uttapap, 2013). In comparison to A-type starches, a more obvious decrease in ΔH value was observed on the B-type after HMT or ANN because of its less compact array of double helices. Furthermore, the phosphate groups that existed in B-type tuber starches could also induce electrostatic repulsion and hinder the arrangement of double helices (Liao et al., 2019).

2.3.4 Starch Digestibility

Starch can be divided into rapidly digestible starch, slowly digestible starch (SDS), and resistant starch (RS), which are hydrolyzed in the simulated *in vitro* system by pancreatin and amyloglucosidase within 20 min, between 20 and 120 minutes, and after 120 minutes, respectively

(Gangoiti et al., 2020). SDS could maintain the stability of postprandial blood sugar, which is thus conducive to preventing or managing the related metabolic syndrome. RS has a prebiotic influence on bifidobacterial in intestines (Fuentes-Zaragoza et al., 2010). As hydrothermal treatments strengthen the starch structures, the enzymatic hydrolysis is supposed to be suppressed (Li et al., 2023). HMT contributed to higher SDS or RS proportions of wheat starch (Chen et al., 2015), high-AM maize starch (Wang et al., 2016), rice starch (Wang et al., 2018b), and millet starch (Babu & Mohan, 2019), as well as many pulse starches, such as mung bean starch (Barua & Srivastav, 2017), chickpea and lentil starches (Chung, Liu, & Hoover, 2010), black and broad bean starches (Chávez-Murillo et al., 2018). This change is largely related to the applied moisture levels, being more obvious at higher values (Babu et al., 2019). HMT also determined an increased SDS or RS of wheat flour (Chen et al., 2015) and rice flour (Kim, Oh, & Chung, 2017). The denaturation of proteins caused by high temperature during HMT could form a layer on starch surfaces which hinder enzymatic absorption and catalysis (Chen et al., 2015). A similar result was accompanied by the application of ANN on sweet potato starch (Trung et al., 2017), grass pea starch (Piecyk et al., 2018), and black bean, broad bean, chickpea, and lentil starches (Chávez-Murillo et al., 2018). Notably, HMT was reported to be more effective in inhibiting starch digestion than ANN because HMT allowed for the formation of more ordered and stable structures (Liu et al., 2016). Digestion behaviors of starches after hydrothermal treatments are also strongly affected by their botanical varieties. High-AM starch presented lower susceptibility to digestion than the low AM one because less AM acts as reinforcing rods between AP (Xu et al., 2020). Starches with a higher degree of surface roughness are more susceptible to enzyme reactions since surface pores provide additional binding sites for enzymatic penetration (Lv et al., 2021). Moreover, the B-type starch composed of hexagonal unit cells is easier to break than the A-type starch consisting of monoclinic unit cells during hydrothermal processing (Hoover, 2010).

2.4 HTS APPLICATIONS

Hydrothermal treatment has been broadly designed for imparting starch and starchy products with preferable characteristics. At present, the applications of HTS mainly concentrated on the food industry. For example, starch granules pregelatinized by hydrothermal treatment are typically used in dry mixes for production of graininess, pulpiness, or smooth pastes when mixed with cold or warm water, such as instant soups (BeMiller et al., 2015).

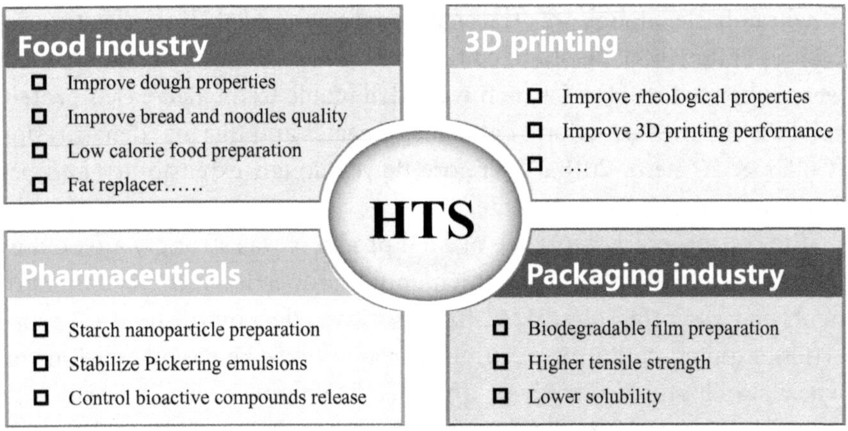

Food industry
- Improve dough properties
- Improve bread and noodles quality
- Low calorie food preparation
- Fat replacer……

3D printing
- Improve rheological properties
- Improve 3D printing performance
- ….

HTS

Pharmaceuticals
- Starch nanoparticle preparation
- Stabilize Pickering emulsions
- Control bioactive compounds release

Packaging industry
- Biodegradable film preparation
- Higher tensile strength
- Lower solubility

FIGURE 2.6 Potential applications of HTS in industries.

HMT- and/or ANN-modified starches are also applied in limited-moisture systems for softness improvements due to the higher water absorption ability, such as starch gels, cereal products, frostings, and toppings. In Figure 2.6, the applications of HMT- and/or ANN-modified starches in the food industry are summarized, and their novel applications in other fields are also discussed.

2.4.1 Food Products

In this part, HMT is mainly mentioned because it is more efficient than ANN in improving product properties. The dough is formed from flour and water by mixing and kneading, in which gluten proteins comprise a continuous network, and starch granules act as a filler, conferring a unique viscoelasticity (Hu et al., 2023). Hence, water availability and gluten and starch characteristics are responsible for dough properties. After HMT, maize starch absorbs more water due to the existence of fissures or hollows on surfaces, which contributes to a higher resistance to mechanical deformation (Miyazaki & Morita, 2005). The mixture of cassava and sodium stearoyl lactylate treated by HMT was incorporated with wheat flour, resulting in higher ratios of dough extension resistance to extensibility and lower AP retrogradation kinetics (Dudu et al., 2020). Increases in storage and loss moduli (G', G'') of dough prepared from HMT-treated wheat–barley mixed flours were obtained, so as to lower tan δ, indicating a highlighted elasticity and a stiffening behavior, which was caused by the higher compactness of starch structures and solubility of barley fibers (Lazaridou, Marinopoulou, & Biliaderis, 2019). The gas-holding ability of wheat–barley dough was thus supposed to augment. However, the viscosities of dough made of mixed teff, chestnut, chickpea, and wheat flours were enhanced by HMT, which was attributable to the decreased protein solubility by the formation of gluten aggregates and thermal denaturation (Collar & Armero, 2018). Therefore, lower dough extensibility and gas retention capacity were presented.

The consumers' acceptance of final products was strongly affected by their texture and volume, in which dough properties and bread components play essential roles. HMT has a positive effect on the textural properties of gluten-containing crumbs through preferable modifications on wheat starch and barley fibers. The specific volume of wheat bread with a substitution of 30% HMT-maize starch was higher (Miyazaki et al., 2005). HMT was responsible for wheat–barley bread with a softer crumb structure (Collar et al., 2018). HMT on rice and corn gluten-free flours improved the composite bread quality, including the volume, softness, and chewiness but lowered the springiness and cohesiveness that separately represented the elasticity and deformation capacity (Bourekoua et al., 2016). The better air encirclement of gluten-free baking products was also explained by higher viscosities because of starch pregelatinization (Fathi et al., 2016). HMT with a high temperature and long duration retarded the short-term and long-term retrogradation through small starch molecules originating from heating degradation (Han et al., 2003; Li et al., 2020b). Similar results of hardness or volumes were reported on cakes prepared by HMT-proso millet flour (Fathi et al., 2016) and cookies prepared with wheat flour and HMT-germinated brown rice flour (Chung, Cho, & Lim, 2014).

For noodle manufacture, higher firmness and elasticity and lower stickiness are desirable. An inhibited swelling power and solubility gave rise to less extent of noodle cooking loss and breakage (Diao et al., 2022). The crystalline arrangements and granular agglomeration of potato starch were accelerated by HMT, and hence the produced noodles were all more acceptable together with less solid loss and breakages (Yang et al., 2022). However, wheat noodle softness was also enhanced by adding HMT-maize starch due to inhibitions on the short-term retrogradation (Ma et al., 2021). In addition, nutritional benefits were brought since the produced noodles had higher contents of RS. Ashwar et al. (2016) reported that when RS was directly incorporated into bread recipes, bread staling was retarded because RS can inhibit AP recrystallization by binding water molecules.

HMT and ANN could also enhance RS levels while maintaining the granular structure. The RS content in high-AM corn starch increased from 18.4% to 43.9% and 28.1% after HMT and ANN, respectively (Brumovsky and Thompson, 2001). RS can be incorporated into foods without affecting their appearance and texture due to its bland taste and white color, thus allowing for the applications as a fat mimetic or to prepare low-calorie food (Lee & Kang, 2024; Jayakody & Hoover, 2008).

2.4.2 Functional Gels and 3D Printing

HMT and/or ANN were applied in waxy maize, tapioca, and rice starches to obtain starch gels with higher water solubility indexes and inhibited retrogradation, which resulted from the inhibited swelling power and pasting behavior (Sun et al., 2021; Włodarczyk-Stasiak et al., 2019). The viscoelastic and mechanical profiles of starch gels are more obviously changed by HMT due to the higher plasticization of starch crystalline and amorphous parts relative to ANN (Marboh & Mahanta, 2021). As a result of great AM molecules being involved in the short-term molecular reorganization, starch gels had a stronger tendency to shear-thinning compared to the untreated group (Liao et al., 2019). The G′, G″, and tan δ (G″/G′) of acorn starch pastes were decreased in the order: native starch>ANN starch>HMT starch (Molavi & Razavi, 2018). Changes in the rheological properties of modified starches are highly dependent on their structural properties. The increase in elastic behavior of HMT-elephant foot yam starch was noticed, which had a positive correlation with the granular size and proportion of short AP chains (Barua et al., 2022). The improved rheological characteristics of starch gels would contribute to improving its 3D printing performance. For example, Zheng et al. (2023) suggested that HMT improved the quality of 3D-printing wheat starch gels by increasing the M_w of AM and decreasing the average M_w.

2.4.3 Other Novel Applications

Some novel applications of HMT and/or ANN-modified starches have been recently developed, and researchers are exploring more potential applications of HTS. A combination of HMT and ANN on rice starch contributed to preparing starch nanoparticles with a superior oil and water absorption performance and water dispersibility, which is useful in stabilizing Pickering emulsions and controlling the release of bioactive compounds (Koh & Liao, 2023). Therefore, HTS might also exhibit great potential in functional foods, pharmaceuticals, cosmetics, etc. HMT has also been used to prepare biodegradable films. The film elaborated with HMT starch presented higher peak force, puncture energy, tensile strength, and lower solubility (Singh, Bawa, Riar, & Saxena, 2009), and HMT had a synergistic improvement on the potato starch nanocomposite film with lignin-sulfonic acid (Chu et al., 2022). They suggested that the film-forming ability of HTS presented a promising future for exploration as a packaging material.

REFERENCE, BIBLIOGRAPHY OR WORKS CITED

Adebowale, K. O., & Olu-Owolabi, B. I. 2011. Effect of heat moisture treatment and annealing on physicochemical properties of red sorghum starch. *African Journal of Biotechnology* 4(9): 928–933.

Ambigaipalan, P., Hoover, R., Donner, E., & Liu, Q. 2014. Starch chain interactions within the amorphous and crystalline domains of pulse starches during heat-moisture treatment at different temperatures and their impact on physicochemical properties. *Food Chemistry* 143: 175–184.

Arns, B., Bartz, J., Radunz, M., Evangelho, J. A. D., Pinto, V. Z., Zavareze, E. D. R., & Dias, A. R. G. 2015. Impact of heat-moisture treatment on rice starch, applied directly in grain paddy rice or in isolated starch. *LWT-Food Science and Technology* 60(2, Pt 1): 708–713.

Ashwar, B. A., Gani, A., Wani, I. A., Shah, A., Masoodi, F. A., & Saxena, D. C. 2016. Production of resistant starch from rice by dual autoclaving-retrogradation treatment: Invitro digestibility, thermal and structural characterization. *Food Hydrocolloids* 56: 108–117.

Babu, S. A., & Mohan, R. J. 2019. Influence of prior pre-treatments on molecular structure and digestibility of succinylated foxtail millet starch. *Food Chemistry* 295: 147–155.

Barua, S., Hanewald, A., Bächle, M., Mezger, M., Srivastav, P. P., & Vilgis, T. A. 2022. Insights into the structural, thermal, crystalline and rheological behavior of various hydrothermally modified elephant foot yam (*Amorphophallus paeoniifolius*) starch. *Food Hydrocolloids* 129: 107672.

Barua, S., & Srivastav, P. 2017. Effect of heat-moisture treatment on resistant starch functional and thermal properties of mung bean (vigna radiate) starch. *Journal of Nutritional Health and Food Engineering* 7(4): 358–363.

BeMiller, J. N. 2011. Pasting, paste, and gel properties of starch-hydrocolloid combinations. *Carbohydrate Polymers* 86(2): 386–423.

BeMiller, J. N., & Huber, K. C. 2015. Physical modification of food starch functionalities. *Annual Review of Food Science and Technology* 6: 19–69.

Bet, C. D., Oliveira, C. S. D., Colman, T. A. D., Marinho, M. T., Lacerda, L. G., Ramos, A. P., & Schnitzler, E. 2018. Organic amaranth starch: A study of its technological properties after heat-moisture treatment. *Food Chemistry* 264: 435–442.

Bharti, I., Singh, S., & Saxena, D. C. 2019. Exploring the influence of heat moisture treatment on physicochemical, pasting, structural and morphological properties of mango kernel starches from Indian cultivars. *LWT-Food Science and Technology* 110: 197–206.

Bourekoua, H., Benatallah, L., Zidoune, M. N., & Rosell, C. M. 2016. Developing gluten free bakery improvers by hydrothermal treatment of rice and corn flours. *LWT-Food Science and Technology* 73: 342–350.

Brumovsky, J. O., & Thompson, D. B. 2001. Production of boiling-stable granular resistant starch by partial acid hydrolysis and hydrothermal treatments of high amylose maize starch. *Cereal Chemistry* 78: 680–689.

Chavez-Murillo, C. E., Orona-Padilla, J. L., & Rosa-Millán, D. L. J. 2019. Physicochemical, functional properties and ATR-FTIR digestion analysis of thermally treated starches isolated from black and bayo beans. *Starch-Stärke* 71 (3–4): 1800250.

Chávez-Murillo, C. E., Veyna-Torres, J. I., Cavazos-Tamez, L. M., Rosa-Millán, D. L. J., & Serna-Saldívar, S. O. 2018. Physicochemical characteristics, ATR-FTIR molecular interactions and in vitro starch and protein digestion of thermally-treated whole pulse flours. *Food Research International* 105: 371–383.

Chen, S., Qin, L., Chen, T., Yu, Q., Chen, Y., Xiao, W. H., Ji, X. Y., & Xie, J. H. 2022. Modification of starch by polysaccharides in pasting, rheology, texture and in vitro digestion: A review. *International Journal of Biological Macromolecules* 207: 81–89.

Chen, X., He, X. W., Fu, X., & Huang, Q. 2015. *In vitro* digestion and physicochemical properties of wheat starch/flour modified by heat-moisture treatment. *Journal of Cereal Science* 63: 109–115.

Chu, T., Shi, J. S., Xia, Y. Z., Wang, H. D., Fan, G. L., & Yang, M. L. 2022. Development of high strength potato starch nanocomposite films with excellent UV-blocking performance: Effect of heat moisture treatment synergistic with ligninsulfonic acid. *Industrial Crops and Products* 187: 115327.

Chung, H. J., Cho, A., & Lim, S. T. 2014. Utilization of germinated and heat-moisture treated brown rices in sugar-snap cookies. *LWT-Food Science and Technology* 57(1): 260–266.

Chung, H. J., Liu, Q., & Hoover, R. 2010. Effect of single and dual hydrothermal treatments on the crystalline structure, thermal properties, and nutritional fractions of pea, lentil, and navy bean starches. *Food Research International* 43(2): 501–508.

Collar, C. 2017. Significance of heat-moisture treatment conditions on the pasting and gelling behaviour of various starch-rich cereal and pseudocereal flours. *Food Science and Technology International* 23(7): 623–636.

Collar, C., & Armero, E. 2018. Value-added of heat moisture treated mixed flours in wheat-based matrices: A functional and nutritional approach. *Food and Bioprocess Technology* 11(8): 1536–1551.

Cordeiro, M. J. M., Veloso, C. M., Santos, L. S., Bonomo, R. C. F., Caliari, M., & Fontan, R. D. C. I. 2018. The impact of heat-moisture treatment on the properties of *Musa paradisiaca* L. starch and optimization of process variables. *Food Technology and Biotechnology* 56(4): 506–515.

Diao, J. J., Tao, Y., Chen, H. S., Zhang, D. J., & Wang, C. Y. 2022. Hydrothermal-induced changes in the gel properties of mung bean proteins and their effect on the cooking quality of developed compound noodles. *Frontiers in Nutrition* 9: 957487.

Dudu, O. E., Ma, Y., Adelekan, A., Oyedeji, A. B., Oyeyinka, S. A., & Ogungbemi, J. W. 2020. Bread-making potential of heat-moisture treated cassava flour-additive complexes. *LWT-Food Science and Technology* 130: 109477.

Falade, K. O., Ibanga-Bamijoko, B., & Ayetigbo, O. E. 2019. Comparing properties of starch and flour of yellow-flesh cassava cultivars and effects of modifications on properties of their starch. *Journal of Food Measurement and Characterization* 13(4): 2581–2593.

Fathi, B., Aalami, M., Kashaninejad, M., & Mahoonak, A. S. 2016. Utilization of heat-moisture treated proso millet flour in production of gluten-free pound cake. *Journal of Food Quality* 39(6): 611–619.

Fuentes-Zaragoza, E., Riquelme-Navarrete, M. J., Sánchez-Zapata, E., & Pérez-Álvarez, J. A. 2010. Resistant starch as functional ingredient: A review. *Food Research International* 43(4): 931–942.

Gangoiti, J., Corwin, S. F., Lamothe, L. M., Vafiadi, C., Hamaker, B. R., & Dijkhuizen, L. 2020. Synthesis of novel α-glucans with potential health benefits through controlled glucose release in the human gastrointestinal tract. *Critical Reviews in Food Science and Nutrition* 60(1): 123–146.

Gunaratne, A., & Hoover, R. 2002. Effect of heat-moisture treatment on the structure and physicochemical properties of tuber and root starches. *Carbohydrate Polymers* 49(4): 425–437.

Han, J. A., BeMiller, J. N., Hamaker, B., & Lim, S. T. 2003. Structural changes of debranched corn starch by aqueous heating and stirring. *Cereal Chemistry* 80(3): 323–328.

Henning, F. G., Schnitzler, E., Demiate, I. M., Lacerda, L. G., Ito, V. C., Malucelli, L. C., & Silva-Carvalho-Filho, M. A. 2019. Fortified rice starches: The role of hydrothermal treatments in zinc entrapment. *Starch-Stärke* 71 (1–2): 1800130.

Hoover, R. 2010. The impact of heat-moisture treatment on molecular structures and properties of starches isolated from different botanical sources. *Critical Reviews in Food Science and Nutrition* 50(9): 835–847.

Hu, X. H., Cheng, L., Hong, Y., Li, Z. F., Li, C. M., & Gu, Z. B. 2023. An extensive review: How starch and gluten impact dough machinability and resultant bread qualities. *Critical Reviews in Food Science and Nutrition* 63(13): 1930–1941.

Huang, T. T., Zhou, D. N., Jin, Z. Y., Xu, X. M., & Chen, H. Q. 2016. Effect of repeated heat-moisture treatments on digestibility, physicochemical and structural properties of sweet potato starch. *Food Hydrocolloids* 54(Pt A): 202–210.

Iuga, M., & Mironeasa, S. 2020. A review of the hydrothermal treatments impact on starch based systems properties. *Critical Reviews in Food Science and Nutrition* 60(22): 3890–3915.

Jacobs, H., & Delcour, J. A. 1998. Hydrothermal modifications of granular starch, with retention of the granular structure: A review. *Journal of Agricultural and Food Chemistry* 46(8): 2895–2905.

Jayakody, L., & Hoover, R. 2008. Effect of annealing on the molecular structure and physicochemical properties of starches from different botanical origins - A review. *Carbohydrate Polymers* 74(3): 691–703.

Ji, N., Ge, S. J., Li, M., Wang, Y. F., Xiong, L., Qiu, L. Z., Bian, X. L., Sun, C. R., & Sun, Q. J. 2019. Effect of annealing on the structural and physicochemical properties of waxy rice starch nanoparticles: Effect of annealing on the properties of starch nanoparticles. *Food Chemistry* 286: 17–21.

Jia, R. Y., Cui, C. L., Gao, L., Qin, Y., Ji, N., Dai, L., Wang, Y. F., Xiong, L., Shi, R., & Sun, Q. J. 2023. A review of starch swelling behavior: Its mechanism, determination methods, influencing factors, and influence on food quality. *Carbohydrate Polymers* 321: 121260.

Kim, J. Y., & Huber, K. C. 2013. Heat-moisture treatment under mildly acidic conditions alters potato starch physicochemical properties and digestibility. *Carbohydrate Polymers* 98(2): 1245–1255.

Kim, M. J., Oh, S. G., & Chung, H. J. 2017. Impact of heat-moisture treatment applied to brown rice flour on the quality and digestibility characteristics of Korean rice cake. *Food Science and Biotechnology* 26(6): 1579–1586.

Koh, Y. C., & Liao, H. J. 2023. Preparation and physicochemical characterization of debranched rice starch nanoparticles from mono- and dual-modification by hydrothermal treatments. *Food Bioscience* 55: 103004.

Kumar, S. R., Tangsrianugul, N., Sriprablom, J., Wongsagonsup, R., Wansuksri, R., & Suphantharika, M. 2023. Effect of heat-moisture treatment on the physicochemical properties and digestibility of proso millet flour and starch. *Carbohydrate Polymers* 307: 120630.

Lan, H., Hoover, R., Jayakody, L., Liu, Q., Donner, E., Baga, M., Asare, E. K., Hucl, P., & Chibbar, R. N. 2008. Impact of annealing on the molecular structure and physicochemical properties of normal, waxy and high amylose bread wheat starches. *Food Chemistry* 111(3): 663–675.

Lawal, O. S. 2005. Studies on the hydrothermal modifications of new cocoyam (*Xanthosoma sagittifolium*) starch. *International Journal of Biological Macromolecules* 37(5): 268–277.

Lazaridou, A., Marinopoulou, A., & Biliaderis, C. G. 2019. Impact of flour particle size and hydrothermal treatment on dough rheology and quality of barley rusks. *Food Hydrocolloids* 87: 561–569.

Lee, I., & Kang, T. Y. 2024. Heat-moisture-treated rice starches with different amylose and moisture contents as stabilizers for nonfat yogurt. *Food Chemistry* 436: 137746.

Li, H., Dhital, S., Flanagan, B. M., Mata, J., Gilbert, E. P., & Gidley, M. J. 2020a. High amylose wheat and maize starches have distinctly different granule organization and annealing behaviour: A key role for chain mobility. *Food Hydrocolloids* 105: 105820.

Li, H. T., Zhang, W. Y., Bao, Y. L., & Dhital, S. 2023. Enhancing enzymatic resistance of starch through strategic application of food physical processing technologies. *Critical Reviews in Food Science and Nutrition* 1–24.

Li, M. N., Zhang, B., Xie, Y., & Chen, H. Q. 2019. Effects of debranching and repeated heat-moisture treatments on structure, physicochemical properties and in vitro digestibility of wheat starch. *Food Chemistry* 294: 440–447.

Li, Q. Q., Liu, S. Y., Obadi, M., Jiang, Y. Y., Zhao, F. F., Jiang, S., & Xu, B. 2020b. The impact of starch degradation induced by pre-gelatinization treatment on the quality of noodles. *Food Chemistry* 302: 125267.

Li, S. L., Ward, R., & Gao, Q. Y. 2011. Effect of heat-moisture treatment on the formation and physicochemical properties of resistant starch from mung bean (Phaseolus radiatus) starch. Food Hydrocolloids 25(7): 1702–1709.

Liao, L. Y., Liu, H. H., Gan, Z. P., & Wu, W. G. 2019. Structural properties of sweet potato starch and its vermicelli quality as affected by heat-moisture treatment. *International Journal of Food Properties* 22(1): 1122–1133.

Liu, C., Hong, J., & Zheng, X. L. 2017. Effect of heat-moisture treatment on morphological, structural and functional characteristics of ball-milled wheat starches. *Starch-Stärke* 69(5–6): 1500141.

Liu, X., Wu, J. H., Xu, J. H., Mao, D. Z., Yang, Y. J., & Wang, Z. W. 2016. The impact of heat-moisture treatment on the molecular structure and physicochemical properties of coix seed starches. *Starch-Stärke* 68(7–8): 662–674.

Loubes, M. A., González, L. C., & Tolaba, M. P. 2018. Pasting behaviour of high impact ball milled rice flours and its correlation with the starch structure. *Journal of Food Science and Technology* 55(8): 2985–2993.

Lv, X. X., Hong, Y., Zhou, Q. W., & Jiang, C. C. 2021. Structural features and digestibility of corn starch with different amylose content. *Frontiers in Nutrition* 8: 692673.

Ma, M. T., Li, Z. J., Yang, F., Wu, H. X., Huang, W. Y., Sui, Z. Q., & Corke, H. 2021. Use of heat-moisture treated maize starch to modify the properties of wheat flour and the quality of noodles. *International Journal of Food Science and Technology* 56(7): 3607–3617.

Ma, Z., & Boye, J. I. 2018. Research advances on structural characterization of resistant starch and its structure-physiological function relationship: A review. *Critical Reviews in Food Science and Nutrition* 58(7): 1059–1083.

Maniglia, B. C., Castanha, N., Le-Bail, P., Le-Bail, A., & Augusto, P. E. D. 2021. Starch modification through environmentally friendly alternatives: A review. *Critical Reviews in Food Science and Nutrition* 61(15): 2482–2505.

Marboh, V., & Mahanta, C. L. 2021. Physicochemical and rheological properties and in vitro digestibility of heat moisture treated and annealed starch of sohphlang (*Flemingia vestita*) tuber. *International Journal of Biological Macromolecules* 168: 486–495.

Miyazaki, M., & Morita, N. 2005. Effect of heat-moisture treated maize starch on the properties of dough and bread. *Food Research International* 38(4): 369–376.

Molavi, H., & Razavi, S. M. A. 2018. Dynamic rheological and textural properties of acorn (*Quercus brantii* lindle.) starch: Effect of single and dual hydrothermal modifications. *Starch-Stärke* 70(11–12): 1800086.

Molavi, H., Razavi, S. M. A., & Farhoosh, R. 2018. Impact of hydrothermal modifications on the physicochemical, morphology, crystallinity, pasting and thermal properties of acorn starch. *Food Chemistry* 245: 385–393.

Mustapha, N. A., Roslen, S. N. H., Gafar, F. S. A., Ibadullah, W. Z. W., & Sukri, R. 2019. Characterization of heat-moisture treated *Dioscorea alata purpurea* flour: Impact of moisture level. *Journal of Food Measurement and Characterization* 13(3): 1636–1644.

Olayinka, O. O., Adebowale, K. O., & Olu-Owolabi, B. I. 2008. Effect of heat-moisture treatment on physicochemical properties of white sorghum starch. *Food Hydrocolloids* 22(2): 225–230.

Olu-owolabi, B. I., Afolabi, T. A., & Adebowale, K. O. 2011. Pasting, thermal, hydration, and functional properties of annealed and heat-moisture treated starch of sword bean (*Canavalia gladiata*). *International Journal of Food Properties* 14(1): 157–174.

Pepe, L. S., Moraes, J., Albano, K. M., Telis, V. R. N., & Franco, C. M. L. 2016. Effect of heat-moisture treatment on the structural, physicochemical, and rheological characteristics of arrowroot starch. *Food Science and Technology International* 22(3): 256–265.

Piecyk, M., Drużyńska, B., Ołtarzewska, A., Wołosiak, R., Worobiej, E., & Ostrowska-Ligęza, E. 2018. Effect of hydrothermal modifications on properties and digestibility of grass pea starch. *International Journal of Biological Macromolecules* 118(Pt B): 2113–2120.

Puncha-arnon, S., & Uttapap, D. 2013. Rice starch vs. rice flour: Differences in their properties when modified by heat-moisture treatment. *Carbohydrate Polymers* 91(1): 85–91.

Ratnayake, W. S., & Jackson, D. S. 2008. Starch gelatinization. In *Advances in Food and Nutrition Research*, 55 (pp. 221–268). London: Elsevier Academic Press. https://doi.org/10.1016/S1043-4526(08)00405-1

Rocha, T. S., Felizardo, S. G., Jane, J. L., & Franco, C. M. L. 2012. Effect of annealing on the semicrystalline structure of normal and waxy corn starches. *Food Hydrocolloids* 29: 93–99.

Ruiz, E., Srikaeo, K., & Revilla, L. S. D. L. 2018. Effects of heat moisture treatment on physicochemical properties and starch digestibility of rice flours differing in amylose content. *Food and Applied Bioscience Journal* 6: 140–153.

Samarakoon, E. R. J., Waduge, R., Liu, Q., Shahidi, F., & Banoub, J. H. 2020. Impact of annealing on the hierarchical structure and physicochemical properties of waxy starches of different botanical origins. *Food Chemistry* 303: 125344.

Shen, H. S., Xu, M. J., Su, C. Y., Zhang, B., Ge, X. Z., Zhang, G. Q., & Li, W. H. 2021. Insights into the relations between the molecular structures and physicochemical properties of normal and waxy wheat B-starch after repeated and continuous annealing. *International Journal of Food Science and Technology* 56(12): 6405–6419.

Shi, M. M., Gao, Q. Y., & Liu, Y. Q. 2018. Corn, potato, and wrinkled pea starches with heat-moisture treatment: Structure and digestibility. *Cereal Chemistry* 95(5): 603–614.

Silva, W. M. F., Biduski, B., Lima, K. O., Pinto, V. Z., Hoffmann, J. F., Vanier, N. L., & Dias, A. R. G. 2017. Starch digestibility and molecular weight distribution of proteins in rice grains subjected to heat-moisture treatment. *Food Chemistry* 219: 260–267.

Sindhu, R., & Khatkar, B. S. 2023. Influence of oxidation, acetylation and hydrothermal treatment on structure and functionality of common buckwheat starch. *International Journal of Biological Macromolecules* 253(Pt 5): 127211.

Singh, G. D., Bawa, A. S., Riar, C. S., & Saxena, D. C. 2009. Influence of heat-moisture treatment and acid modifications on physicochemical, rheological, thermal and morphological characteristics of Indian water chestnut (*Trapa natans*) starch and its application in biodegradable films. *Starch-Stärke* 61(9): 503–513.

Siwatch, M., Yadav, R. B., & Yadav, B. S. 2022. Annealing and heat-moisture treatment of amaranth starch: Effect on structural, pasting, and rheological properties. *Journal of Food Measurement and Characterization* 16(3): 2323–2334.

Sun, L. T., Xu, Z. K., Song, L. L., Ma, M. T., Zhang, C. C., Chen, X. J., Xu, X. M., Sui, Z. Q., & Corke, H. 2021. Removal of starch granule associated proteins alters the physicochemical properties of annealed rice starches. *International Journal of Biological Macromolecules* 185: 412–418.

Sun, Q. J., Zhu, X. L., Si, F. M., & Xiong, L. 2015. Effect of acid hydrolysis combined with heat moisture treatment on structure and physicochemical properties of corn starch. *Journal of Food Science and Technology* 52(1): 375–382.

Suriya, M., Reddy, C. K., & Haripriya, S. 2019. Functional and thermal behaviors of heat-moisture treated elephant foot yam starch. *International Journal of Biological Macromolecules* 137: 783–789.

Tester, R. F., & Debon, S. J. J. 2000. Annealing of starch: A review. *International Journal of Biological Macromolecules* 27(1): 1–12.

Tester, R. F., Debon, S. J. J., & Sommerville, M. D. 2000. Annealing of maize starch. *Carbohydrate Polymers* 42(3): 287–299.

Trung, P. T. B., Ngoc, L. B. B., Hoa, P. N., Tien, N. N. T., & Hung, P. V. 2017. Impact of heat-moisture and annealing treatments on physicochemical properties and digestibility of starches from different colored sweet potato varieties. *International Journal of Biological Macromolecules* 105(Pt 1): 1071–1078.

Varatharajan, V., Hoover, R., Li, J., Vasanthan, T., Nantanga, K. K. M., Seetharaman, K., Liu, Q., Donner, E., Jaiswal, S., & Chibbar, R. N. 2011. Impact of structural changes due to heat-moisture treatment at different temperatures on the susceptibility of normal and waxy potato starches towards hydrolysis by porcine pancreatic alpha amylase. *Food Research International* 44(9): 2594–2606.

Vermeylen, R., Goderis, B., & Delcour, J. A. 2006. An X-ray study of hydrothermally treated potato starch. *Carbohydrate Polymers* 64(2): 364–375.

Wang, H. W., Liu, Y. F., Chen, L., Li, X. X., Wang, J., & Xie, F. W. 2018a. Insights into the multi-scale structure and digestibility of heat-moisture treated rice starch. *Food Chemistry* 242: 323–329.

Wang, H. W., Zhang, B. J., Chen, L., & Li, X. X. 2016. Understanding the structure and digestibility of heat-moisture treated starch. *International Journal of Biological Macromolecules* 88: 1–8.

Wang, L., Zhang, C. N., Chen, Z. X., Wang, X. P., Wang, K., Li, Y. F., Wang, R., Luo, X. H., Li, Y. N., & Li, J. 2018b. Effect of annealing on the physico-chemical properties of rice starch and the quality of rice noodles. *Journal of Cereal Science* 84: 125–131.

Włodarczyk-Stasiak, M., Mazurek, A., Jamroz, J., Pikus, S., & Kowalski, R. 2019. Physicochemical properties and structure of hydrothermally modified starches. *Food Hydrocolloids* 95: 88–97.

Xu, J. C., Chen, L., Guo, X. B., Liang, Y., & Xie, F. W. 2020. Understanding the multi-scale structure and digestibility of different waxy maize starches. *International Journal of Biological Macromolecules* 144: 252–258.

Yang, S., Dhital, S., Zhang, M. N., Wang, J., & Chen, Z. G. 2022. Structural, gelatinization, and rheological properties of heat-moisture treated potato starch with added salt and its application in potato starch noodles. *Food Hydrocolloids* 131: 107802.

Zavareze, E. d. R., & Dias, A. R. G. 2011. Impact of heat-moisture treatment and annealing in starches: A review. *Carbohydrate Polymers* 83(2): 317–328.

Zeng, F., Ma, F., Kong, F. S., Gao, Q. Y., & Yu, S. J. 2015. Physicochemical properties and digestibility of hydrothermally treated waxy rice starch. *Food Chemistry* 172: 92–98.

Zheng, L. Y., Zhang, Q. R., Yu, X. Z., Luo, X. H., & Jiang, H. 2023. Effect of annealing and heat-moisture pretreatment on the quality of 3D-printed wheat starch gels. *Innovative Food Science and Emerging Technologies* 84: 103274.

CHAPTER 3

Ultrahigh Pressure Treatment of Starch and Starchy Materials

Qisya Izanti Binti Mohammad Amirul Mursyid and Dandan Li

3.1 INTRODUCTION

Ultrahigh pressure treatment (UHPT) stands out as a cutting-edge and non-thermal technological advancement that holds immense significance for the processing of starch and starchy materials (Castro et al., 2020; Kim, Kim, & Baik, 2012). In stark contrast to conventional methods reliant on heat or chemical additives, UHPT introduces a new approach by subjecting materials to exceptionally elevated pressure, thereby resulting in distinct transformations in their structural, physical, and chemical characteristics. To our knowledge, the first study concerning UHPT-modifying starch was published by Thevelein et al. (1981), who reported that high pressure could dramatically impact the gelatinization temperature of potato starch. Then, in 1982, high pressure was found to induce the gelatinization of potato, wheat, and smooth pea starches below gelatinization temperatures, which triggered a great interest in using UHPT as a gelatinization and/or physical modification method to improve the physicochemical and functional properties of starch (Muhr & Blanshard, 1982; Muhr, Wetton, & Blanshard, 1982). Over the past two decades, there has been a notable surge in the

DOI: 10.1201/9781003493594-3

exploration of high-pressure technology as an alternative to conventional heat treatment for starch processing, and the effect of UHPT parameters such as pressure, temperature, pressure holding time, starch concentration, and starch source was investigated to understand the mechanism and possible applications of pressure-modified starch granules (Castro et al., 2020; Stute et al., 1996). More recently, studies have been performed on the application of UHPT in facilitating the chemical modification of starches in a non-thermal state and improving the physicochemical properties of starch derivatives (Kim et al., 2012; Wang et al., 2023a, b). In Figure 3.1, the timelines of some important events held in the field of UHPT-modified starches are illustrated.

What sets UHPT apart is its capacity to modify starch properties without resorting to additives or subjecting the materials to high temperatures (Castro et al., 2020). This characteristic positions UHPT as a highly promising technology in the realm of clean-label products, aligning seamlessly with the industry's commitment to transparency and simplicity in ingredient lists. UHPT represents a cutting-edge and innovative approach to the treatment of starch and starchy materials. Its exceptional ability to induce gelatinization, alter textures, and preserve nutritional content underscores its versatility, making it a technology poised to revolutionize the landscape of starch processing and product

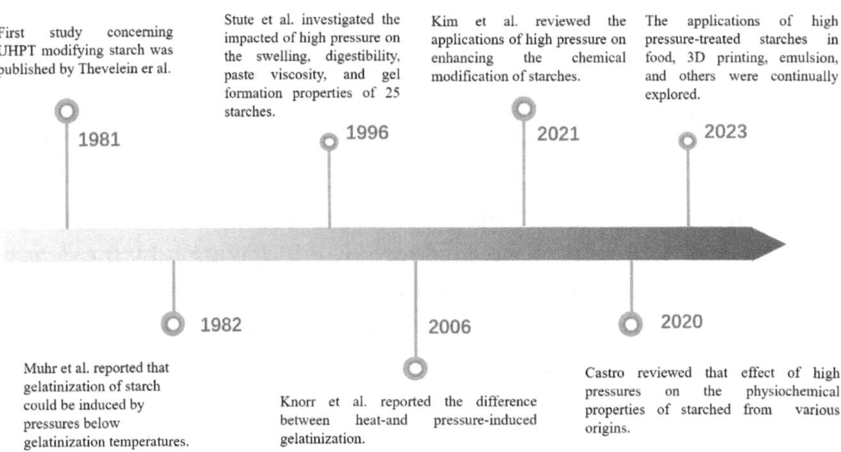

FIGURE 3.1 Timeline of some important events held in the field of high-pressure-modified starches.

innovation (Castro et al., 2020; Gulzar et al., 2023). In the upcoming discussion, we will delve into the fundamental principles that govern UHPT, exploring the structural changes that occur in starches subjected to this treatment, and examining the diverse applications that this technology holds within the realm of food and non-food processing. This investigation promises to shed light on the transformative potential of UHPT and its role in shaping the future of sustainable and minimally processed starchy products.

3.2 PRINCIPLE OF ULTRAHIGH PRESSURE ACTION

3.2.1 Static Pressure

The principle of static pressure action serves as a foundational concept in both fluid mechanics and thermodynamics. It specifically addresses the behaviour of fluids, encompassing both liquids and gases, when subjected to pressure in a state of static equilibrium. In this context, static equilibrium implies that the fluid is not in motion but rather in a stable, stationary state. The term "static pressure" is synonymous with pressure, particularly when used in connection with the Bernoulli equation, aimed at distinguishing it from dynamic pressure.

The Bernoulli equation itself is a fundamental expression in fluid mechanics, articulating that the sum of the flow, kinetic, and potential energies of a fluid particle along a streamline remains constant. This equation sheds light on the intricate interplay between different forms of energy within a fluid system. In essence, the Bernoulli equation reveals that the kinetic and potential energies of the fluid are interchangeable with flow energy, and vice versa, as the fluid undergoes motion. This dynamic interconversion is integral to understanding the changes in pressure that occur during the flow of the fluid. To make this phenomenon more perceptible, the Bernoulli equation can be enhanced by multiplying it by the density (denoted as ρ). This augmentation amplifies our ability to visualize and comprehend the intricate relationship between pressure, energy, and the dynamic behaviour of fluids as elucidated by the principle of static pressure action (Cengel & Cimbala, 2006).

$$P+\rho\frac{V^2}{2}+\rho gz = constant\ (along\ a\ streamline) \tag{3.1}$$

The symbol P in this context represents static pressure, a term denoting the actual thermodynamic pressure of the fluid. It is noteworthy that static pressure, in its essence, does not encompass any dynamic effects. The static pressure aligns with the conventional understanding of pressure as utilized in thermodynamics and property tables. It serves as a measure of the thermodynamic state of the fluid, capturing the equilibrium pressure when the fluid is not in motion (Cengel & Cimbala, 2006).

Although static pressures are evidently mentioned in Bernoulli's principle, it is worth mentioning that Bernoulli's principle is not applied in UHPTs at which instead of investigating aerodynamics or fluid flow in pipes, UHPTs are a direct application of high pressures. However, grasping the fundamentals of static pressure is vital in understanding the parameters that govern the UHPT process. Instead, Le Chatelier's principle, which pertains to changes in temperature, pressure, and volume in a system at equilibrium, is more relevant. When a force is applied, stimulating structural changes and modifications associated with a reduced volume, a new equilibrium is established. The efficiency of UHPT treatment can be influenced by various parameters, including adiabatic heating, the combination of treatment time and temperature, pressure levels, time and pressure combinations, decompression time, intrinsic product factors such as composition, water activity, pH, packaging material, as well as extrinsic parameters during processing, storage, and transportation (Woldemariam & Emire, 2019).

Static pressures, as described here, will later find a correlation with dynamic pressures in Section 3.2.2, specifically in a concept referred to as stagnation pressure. Stagnation pressure introduces a dynamic perspective by considering the impact of motion on the fluid, demonstrating how static and dynamic pressures interrelate. This correlation enhances our comprehension of the fluid's behaviour and its thermodynamic characteristics under varying conditions, providing a comprehensive understanding of the fluid's pressure dynamics. UHPTs employ both static pressures and dynamic pressures in treating starches to induce specific changes in the materials treated.

3.2.2 Dynamic Pressure

The term "dynamic pressure action" encapsulates the effects and principles linked to the dynamic application of pressure within the realm of fluid dynamics. In the domain of fluid mechanics, dynamic pressure constitutes a crucial component of the overall pressure exerted by a fluid in motion, closely tied to the kinetic energy exhibited by the fluid particles engaged in dynamic activity. Like static pressure action, dynamic pressure is present within the framework of the Bernoulli equation, and again, with no direct relation to UHPT. The fundamental understanding of dynamic pressure action and static pressure action is required to fully grasp its roles in UHPT.

As represented in Eq. (3.1), the expression $\frac{V^2}{2}$ is the fluid density and V is the velocity of the fluid, representing the dynamic pressure. This term signifies the pressure increase that occurs when a fluid in motion is isentropically decelerated or brought to a stop. The concept of dynamic pressure is pivotal in fluid dynamics, providing insights into the kinetic energy associated with the movement of fluid particles. In practical terms, when the fluid comes to a halt isentropically, the kinetic energy of the fluid is transformed into an increase in pressure, which is precisely what the dynamic pressure term $\frac{V^2}{2}$ quantifies. This relationship highlights the dynamic nature of fluid behaviour and the consequential alterations in pressure that accompany changes in fluid velocity (Cengel & Cimbala, 2006).

As mentioned in Section 3.2.1, static and dynamic pressures are correlated under a concept known as stagnation pressure at which the sum of static and dynamic pressures yields the stagnation pressure, expressed by the following equation:

$$P_{stag} = P + \rho\frac{V^2}{2}\,(kPa) \tag{3.2}$$

Stagnation pressure is emblematic of the pressure at a specific point where the fluid undergoes isentropic deceleration and comes to a complete stop. Upon measuring static and stagnation pressures at a designated location, it becomes feasible to calculate the fluid velocity at that precise point through appropriate relationships as represented in Eq. (3.3) (Cengel & Cimbala, 2006).

$$v = \sqrt{\frac{2(P_{stag} - P)}{\rho}} \tag{3.3}$$

This comprehensive consideration of pressure allows for a deeper understanding of fluid behaviour and aids in the determination of key parameters, showcasing the interconnected nature of static and dynamic pressures in UHPT.

3.3 STRUCTURAL CHARACTERISTICS OF PRESSURE-TREATED STARCH (PTS)

3.3.1 Comparison of Heat- and Pressure-Induced Gelatinization of Starch

Upon subjecting starch to heating in an excess quantity of water, the transformative process of gelatinization occurs (Ratnayake & Jackson, 2008). During gelatinization, the molecular and crystalline order within the starch granule undergoes a disruption, which leads to the absorption of water, resulting in swelling and eventual fragmentation of the granules. Depending on the specific conditions and type of starch employed, leaching of amylose and amylopectin – the two primary starch polymers – may also occur. The consequential gelatinization process induces a remarkable shift in the rheological properties of the starch suspension, marking a vital transformation in its overall behaviour.

UHPT of starch was commonly conducted to starch–water suspensions, with a pressure ranging from 100 to 600 MPa, a treatment temperature of room temperature (about 20°C–30°C), and a duration of 2–30 minutes (Castro et al., 2020; Gulzar et al., 2023). Extreme treatments involving elevated temperature and pressure can also induce the gelatinization of starch granules in starch–water suspensions (Bauer & Knorr, 2005). Gelatinization is characterized as an irreversible phase transition of starch granules, shifting from an ordered to a disordered state, particularly in the presence of excess water (Douzals et al., 1998). It is noteworthy, however, that pressure-induced gelatinization is deemed notably distinct from its heat-induced counterpart (Kim et al., 2012; Knorr, Heinz, & Buckow, 2006). Initially, the hydration and swelling of amorphous regions within starch granules were facilitated by an increasing pressure level. Next, with the further increase of pressure, the amorphous lamellae within crystalline regions of starch granules start to hydrate and swell, leading to the smectic arrangements of amylopectin double helices. Then, water molecules penetrate into the smectic structure, which then induces the dissociation of amylopectin clusters and unwinding of amylopectin double helices. Unlike traditional heating, the dissociation of crystallite and the unwinding of amylopectin double helix could be restricted by UHPT (Figure 3.2). As a result, gelatinization induced by high pressure undergoes incomplete disintegration of crystallinities within starch granules, which consequently caused different changes in the physicochemical properties

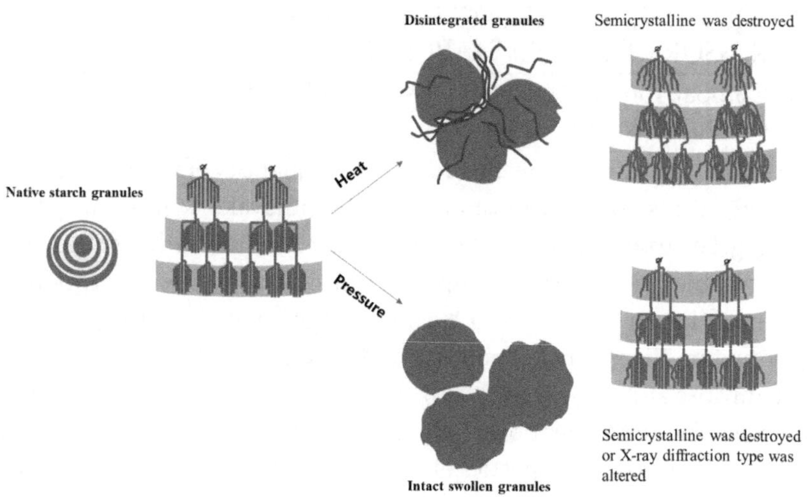

FIGURE 3.2 Comparison of heat- and pressure-induced gelatinization of starch granules (final stage).

(Kim et al., 2012). In Sections 3.3.1–3.3.4, the effects of UHPT on the granule morphology, particle size, birefringence, and semicrystalline are discussed.

3.3.2 Granule Morphology

The UHPT-induced changes in granule morphology of starch from various origins, such as waxy wheat, lotus seed, sorghum, buckwheat, mung bean, rice, proso millet, pea, chestnut, sorghum, and potato, have been investigated (Castro et al., 2020; Stute et al., 1996). Low pressure would not affect the granule morphology, particle size, and birefringence, while high pressure destroyed the granules and caused the gelatinization of starch when dispersed in adequate water. For example, sorghum starch granules generally have a size range of 2–30 μm, notably with polygonal and smaller granules in the corneous endosperm and larger, more rounded granules in the floury endosperm. Native sorghum starch, when observed under SEM as shown in Figure 3.3 (Vallons & Arendt, 2009), displayed these characteristic shapes. However, during the pressure process, a structural transformation occurred. Initially, the granules collapsed into a "doughnut-shaped" configuration at low pressures. With more intense pressure treatment, a heightened level of deformation was observed in the granules. Despite these alterations, treatment at 600 MPa

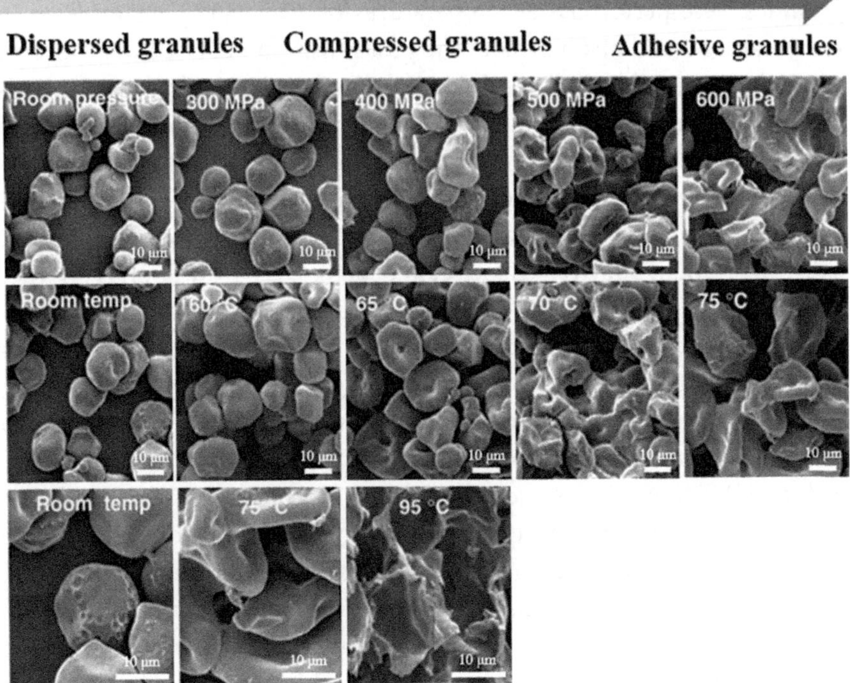

FIGURE 3.3 Scanning microscope images of sorghum starch at various pressures and temperatures and the changes in the particle size with pressures.

(equivalent to 100% gelatinization degree) revealed a noteworthy outcome at which a significant portion of granules retained some degree of integrity. This retention of granular structure is deemed typical in the context of pressure gelatinization, which differs from heat-induced gelatinization. For heat-induced gelatinization, at 95°C, starch granules lost their native shape and formed a viscous paste (Figure 3.3). This difference might be caused by that UHPT could restrict the dissociation of crystallite and the unwinding of amylopectin double helix, which was explained by Kim et al. (2012) and Knorr et al. (2006). Notably, the complete loss of granular integrity could also sometimes be achieved by high pressures. For example, Hu et al. (2017) found that the granular structure of waxy wheat starch treated at 600 MPa was dramatically destroyed and a leaf-like shape appeared. The underlying mechanism for pressure-induced gelatinization still needs to be further explored.

UHPT temperature and duration would also affect the morphology of starch granules (Kim et al., 2012). Although most UHPTs of starch granules were performed at temperatures of 20°C–30°C, an increase in temperature could also promote the gelatinization of starch dispersed in water (Bauer et al., 2005). At higher temperatures, the hydrogen bonds among the amylopectin double helices were weakened, thus allowing the rapid penetration of water molecules to starch crystallites at lower pressure levels, typically below 300 MPa. Notably, the impact of temperature on pressure-induced gelatinization became minor at high pressures, typically higher than 500 MPa (Bauer et al., 2005). UHPT duration is another key factor determining the UHPT-induced destruction of starch granules. At a given pressure and temperature, more extensive destruction of granular morphology was observed with prolonged duration (Buckow, Heinz, & Knorr, 2007). Moreover, the starch concentration and starch structure, especially the semicrystalline type, would also dramatically affect UHPT-induced destruction of starch morphology. For example, an increase in starch concentration from 5% to 80% (w/w) would reduce the gelatinization extent of starch granules (Baks et al., 2008; Kawai, Fukami, & Yamamoto, 2007a, b). Compared with A-type, B-type starch seems to be more resistant to UHPT, which was attributed to the differences in the amylopectin branch-chain structures and packing compressibility (Bauer et al., 2005).

3.3.3 Particle Size

During UHPT, starch granules became tighter at low pressures, while the granules start to stick together to form big aggregates at high pressures. Thereafter, the particle size was supposed to be increased by pressures (Castro et al., 2020). The particle size distribution of pea starch was monomodal with an agglomeration of larger particles at higher UHPT pressures, and an apparent increase in the volume mean diameter and area mean diameter was observed (Leite et al., 2017). Pressure-treated lentil starches exhibited distinct bimodal particle size distribution, and the curve tended to shift to a higher value after 600 MPa treatment, with the overall particle size increasing from 196 to 207 μm (Ahmed et al., 2016). An increase in particle size also happens for other starches when the pressure exceeds 450 MPa (Guo et al., 2015). The increased particle size was confirmed to be caused by the gelatinization. When the UHPT of pea starch dispersed in water and ethanol, the particles dispersed in water showed a significantly larger mean diameter than those in ethanol, which validated the importance of water in promoting gelatinization (Leite et al., 2017). Notably, when conducted under excessive pressure, slight degradation might occur, which could decrease the particle size of starch granules. In addition to pressure, the treatment duration, treatment temperature, and starch origins were also responsible for the alteration in starch particle size.

3.3.4 Birefringence

Starch granules possess birefringence properties, which could be characterized by the presence of a Maltese cross. The crosses were formed due to the radial orientation of the double helices of amylopectin within crystallinities. During UHPT, water diffuses into the crystalline regions, which then disrupts the hydrogen bond network among amylopectin chains. Thereafter, a loss of birefringence in starch granules is observed (Deng et al., 2014). At low pressures, typically below 300 MPa, no special changes in the Maltese cross were observed. However, at moderate pressures (400–500 MPa), a loss of birefringence and Maltese cross happened. When the pressure was high enough (~600 MPa), starch was completely gelatinized, and thus an entirely disappearance of birefringence occurred. For example, as depicted in Figure 3.4 (Vallons et al., 2009), the tested sorghum starch granules displayed characteristic shapes and birefringence patterns at different temperatures and pressures at which the number of granules exhibiting the Maltese cross pattern decreased with increased pressure beyond 300 MPa and increased temperature beyond 60°C. At conditions corresponding to 400 MPa and 65°C, a pronounced loss in birefringence was observed. Remarkably, all granules lost their Maltese cross signature after undergoing treatment with 600 MPa or 75°C. These findings underscore the significant structural alterations experienced by sorghum starch granules under varying pressure and temperature conditions. A similar phenomenon was observed for other starches including lotus seed (Guo et al., 2015), red adzuki bean (Li et al., 2015), rice (Deng et al., 2014), etc.

3.3.5 Semicrystalline

Starch is predominantly differentiated by their polymorphs, including A-type (cereal starches), B-type (tuber, root, high-amylose cereal starches, and retrograded starches), and C-type (legume starches) crystalline forms (Yashini et al., 2022). The impact of high pressure on the semicrystalline structure of starch with different X-ray types is shown in Figure 3.5. With increasing pressure, temperature, and duration, the intensity of

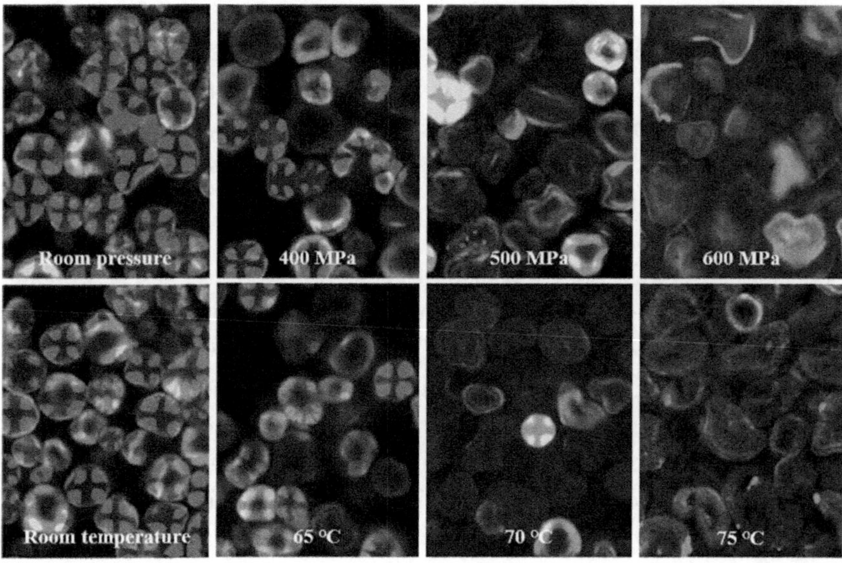

FIGURE 3.4 Laser confocal microscope images of sorghum starch subjected to various pressures and temperatures.

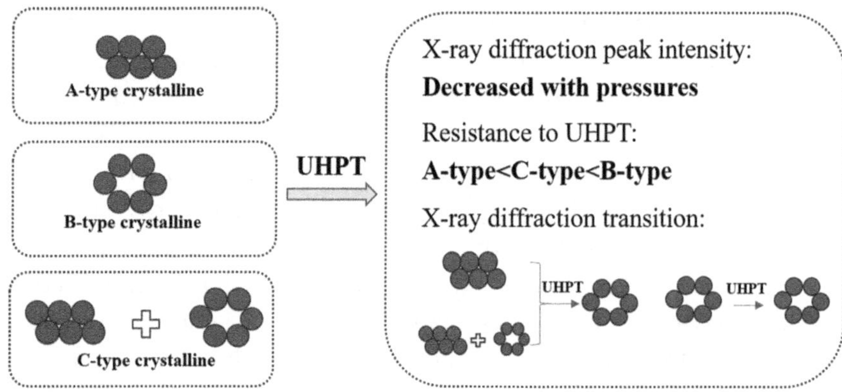

FIGURE 3.5 Changes in semicrystalline structure of starch granules subjected to pressure.

X-ray diffraction peaks got weaker, due to the disruption of starch crystallinities. Furthermore, A-type starches are typically the most sensitive to UHPT, B-type are more resistant to pressure, while C-type are located in between B- and A-type starches (Vallons et al., 2009). For example, wheat starch with an A-type was confirmed to be the most sensitive to pressure, potato starch showed the highest resistance, while tapioca starch started to gelatinize at higher pressures than wheat starch but lower than potato starch, due to the C-type crystalline structure of tapioca starch (Ezaki & Hayashi, 1992). The transformation of X-ray polymorphic form could also be induced by HPT. A- and C-types tend towards B-type after gelatinization caused by HPT, while B-type maintains their native X-ray pattern. Deng et al. (2014), Hu et al. (2017), Li et al. (2012), Li et al. (2018), and Liu et al. (2016a, b, c) reported the transition of X-ray pattern from A-type to B-type for rice, waxy wheat, proso millet, sorghum, Tartary buckwheat, and common buckwheat starches.

3.4 PRESSURE-TREATED STARCH (PTS) APPLICATION

The structural changes induced by UHPT would further affect the physicochemical properties of starch such as thermal stability, rheological properties, pasting properties, gel clarity and transparency, retrogradation properties, and *in vitro* digestibility (Castro et al., 2020), making it a promising technology across pharmaceutical, food, nutraceutical, health, and related industries. This section aims to elucidate these applications in detail (Figure 3.6).

UHPT has proven effective in mitigating the occurrence of caramelization and Maillard reactions, offering distinct advantages over traditional thermal methods. The preservation of original flavours, colours, nutrients,

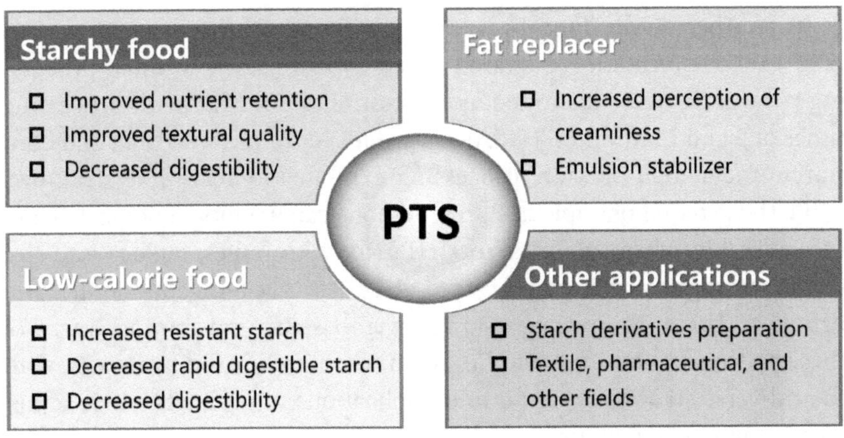

FIGURE 3.6 Applications of pressure-treated starch (PTS) in food and other industries.

and overall food quality sets UHPT apart. Extensive research has delved into the inactivation of microorganisms across diverse products, including seafood, meat, eggs, dairy, and beverages. UHPT achieves microbial load inhibition by inducing structural changes in cell membranes and triggering chemical reactions that impede microbial growth (Allai et al., 2023). The versatility of UHPT technology extends to various promising food applications, contributing to nutritional enhancement, prolonged shelf life with minimal preservative use, reduced food spoilage, preservation of freshness, and retention of sensory characteristics (Khaliq et al., 2021).

In the research conducted by Xia et al. (2017) on brown rice, the optimal processing parameters were identified as 30 MPa for a duration of 10 minutes. The study revealed notable enhancements in the *in vitro* bio-accessibility of calcium and copper by 12.59%–52.17% and 2.87%–23.06%, respectively, following high-pressure processing. However, the bio-accessible iron exhibited a decrease. Furthermore, high pressure demonstrated a significant improvement in individual amino acids, particularly indispensable amino acids and gamma-aminobutyric acid. The study also highlighted increased bio-accessible total antioxidant activities and enhanced starch resistance to enzymatic hydrolysis. It is worth noting that despite these improvements, germination was found to substantially increase starch digestibility. Atomic force microscopy characterization indicated a noticeable structural change in the bran fraction at pressures above 300 MPa. Significantly, free amino acid content, bio-accessibility, and resistant starch were all markedly enhanced with an increase in treatment pressure.

In another study done by Larrea-Wachtendorff, Sousa, and Ferrari (2021) on the formation of starch-based hydrogels, the optimal processing parameters were identified as a pressure of 600 MPa with processing times of 5 and 15 minutes. Hydrogels produced from rice, wheat, and corn starch under high-pressure processing exhibited a cream-like structure, while those based on tapioca starch demonstrated a more compact structure. Notably, when subjected to UHPT for 15 minutes, tapioca and rice starch hydrogels displayed increased viscosity, G′ (storage modulus), and firmness. This outcome suggests an overall structural reinforcement of these hydrogels. The study implies that these starch-based hydrogels, with their diverse structures, could find applications in products where either creamy or gummy textures are desirable.

High-pressure treatment could also promote the formation of resistant and slowly digestible starch, thus exhibiting great potentials in low-calorie food production (Kim, Ye, & Baik, 2023). During high-pressure treatment, the starch structure, especially the interactions between amylose and amylopectin chains, was altered, thus decreasing the susceptibility of digestion enzymes to modified starch. Liu et al. (2016a, b, c) found that amylose content and crystallinity of UHPT-modified starches were superior to the native starch. As a result, the contents of resistant and slow digestible starch for Tartary buckwheat, common buckwheat, and sorghum starches were increased by UHPT. Moreover, UHPT in combination with starch retrogradation was applied to reduce the *in vitro* gastro-small intestinal digestion of potato starch. Colussi et al. (2018) found a significant reduction of 10%–15% in the hydrolysis of starch modified by six cycles for 10 minutes at 400 MPa with retrogradation in relation to the native and only pressure-processed starch. Overall, high pressure is a good technology to obtain starches with increased potential health benefits.

In the realm of fat replacers, it is crucial to note the fundamental distinction between two main categories which are fat substitutes, which originate from lipids, and fat mimetics, deeply rooted in carbohydrates and proteins. Specifically, when we dive into carbohydrate-based fat mimetics, such as starches found in everyday grains like wheat, corn, and waxy corn, we are able to discover their widespread use across a wide spectrum of food formulations. These starches, widely utilized for their economic benefits and versatility, play an indispensable role as thickeners, gelling agents, and stabilizers in various culinary applications. However, despite their ubiquity, native starches encounter inherent challenges when employed as fat replacers, grappling with issues like insufficient creaminess, adhesiveness, and a propensity for retrogradation (Heydari & Razavi, 2021). UHPT method stands out as a cornerstone for altering the properties of starches, particularly in the realm of crafting fat substitutes for emulsions with reduced fat content (Heydari & Razavi, 2021; Heydari, Razavi, & Farahnaky, 2021). The resulting PTSs boast modified crystalline structures, optimal zeta potential, a lower viscosity, and a diminished retrogradation tendency, thereby influencing how fats are perceived during oral consumption. These starches exhibit pronounced cohesiveness, consistency, a smaller particle size, and higher thixotropy, all of which contribute to an enhanced release of flavour, a velvety mouthfeel, and an increased perception of creaminess during consumption. Moreover, their heightened water absorption capacity and enhanced thermal resistance play pivotal roles in elevating the creaminess and stability of emulsions over time. In summation, the introduction of PTSs stands as a finely tuned alternative, poised to address the multifaceted requirements for fat substitution in reduced-fat emulsions. The journey from inherent challenges to promising solutions reflects the dynamic and ever-evolving landscape of food science and technology.

Nowadays, the applications of UHPT-modified starches in food and other industries are continuously explored. For example, Kim et al. (2012) reviewed the functional roles of UHPT in the acid-hydrolysis, hydroxypropylation, acetylation, and cross-linking modification of starch granules, and the physicochemical properties of the starch derivatives prepared with UHPT were discussed. These starch derivatives might exhibit various potential applications in food and other industries such as 3D printing, textile, pharmaceutical, and so on.

REFERENCE, BIBLIOGRAPHY OR WORKS CITED

Ahmed, J., Thomas, L., Taher, A., & Joseph, A. 2016. Impact of high pressure treatment on functional, rheological, pasting, and structural properties of lentil starch dispersions. *Carbohydrate Polymers* 152: 639–647.

Allai, F. M., Azad, Z. R. A. A., Mir, N. A., & Gul, K. 2023. Recent advances in non-thermal processing technologies for enhancing shelf life and improving food safety. *Applied Food Research* 3(1): 100258.

Baks, T., Bruins, M. E., Janssen, A. E. M., & Boom, R. M. 2008. Effect of pressure and temperature on the gelatinization of starch at various starch concentrations. *Biomacromolecules* 9(1): 296–304.

Bauer, B. A., & Knorr, D. 2005. The impact of pressure, temperature and treatment time on starches: Pressure-induced starch gelatinisation as pressure time temperature indicator for high hydrostatic pressure processing. *Journal of Food Engineering* 68(3): 329–334.

Buckow, R., Heinz, V., & Knorr, D. 2007. High pressure phase transition kinetics of maize starch. *Journal of Food Engineering* 81(2): 469–475.

Castro, L. M. G., Alexandre, E. M. C., Saraiva, J. A., & Pintado, M. 2020. Impact of high pressure on starch properties: A review. *Food Hydrocolloids* 106: 105877.

Cengel, Y. A., & Cimbala, J. M. 2006. Fliud mechanics Anderson: Computational fluid dynamics: The basics with applications. In *McGraw-Hill Series in Mechanical Engineering*, ed. S. Jeans, 35–58. New York, the McGraw-Hill companies.

Colussi, R., Kaur, L., Zavareze, E. d. R., Dias, A. R. G., Stewart, R. B., & Singh, J. 2018. High pressure processing and retrogradation of potato starch: Influence on functional properties and gastro-small intestinal digestion in vitro. *Food Hydrocolloids* 75: 131–137.

Deng, Y., Jin, Y., Luo, Y., Zhong, Y., Yue, J., Song, X., & Zhao, Y. 2014. Impact of continuous or cycle high hydrostatic pressure on the ultrastructure and digestibility of rice starch granules. *Journal of Cereal Science* 60(2): 302–310.

Douzals, J. P., Perrier Cornet, J. M., Gervais, P., & Coquille, J. C. 1998. High-pressure gelatinization of wheat starch and properties of pressure-induced gels. *Journal of Agricultural and Food Chemistry* 46(12): 4824–4829.

Ezaki, S, Hayashi, R. 1992. High pressure effects on starch: Structural change and retrogradation. *Colloques-Institut National De La Sante Et De La Recherche Medicale Colloques Et Seminaires* 1992: 163–163.

Gulzar, S., Narciso, J. O., Elez-Martínez, P., Martín-Belloso, O., & Soliva-Fortuny, R. 2023. Recent developments in the application of novel technologies for the modification of starch in light of 3D food printing. *Current Opinion in Food Science* 52: 101067.

Guo, Z., Zeng, S., Lu, X., Zhou, M., Zheng, M., & Zheng, B. 2015. Structural and physicochemical properties of lotus seed starch treated with ultra-high pressure. *Food Chemistry* 186: 223–230.

Heydari, A., & Razavi, S. M. A. 2021. Evaluating high pressure-treated corn and waxy corn starches as novel fat replacers in model low-fat O/W emulsions: A physical and rheological study. *International Journal of Biological Macromolecules* 184: 393–404.

Heydari, A., Razavi, S. M. A., & Farahnaky, A. 2021. Effect of high pressure-treated wheat starch as a fat replacer on the physical and rheological properties of reduced-fat O/W emulsions. *Innovative Food Science & Emerging Technologies* 70: 102702.

Hu, X. P., Zhang, B., Jin, Z. Y., Xu, X. M., & Chen, H. Q. 2017. Effect of high hydrostatic pressure and retrogradation treatments on structural and physicochemical properties of waxy wheat starch. *Food Chemistry* 232: 560–565.

Kawai, K., Fukami, K., & Yamamoto, K. (2007a). Effects of treatment pressure, holding time, and starch content on gelatinization and retrogradation properties of potato starch-water mixtures treated with high hydrostatic pressure. *Carbohydrate Polymers* 69(3): 590–596.

Kawai, K., Fukami, K., & Yamamoto, K. 2007b. State diagram of potato starch-water mixtures treated with high hydrostatic pressure. *Carbohydrate Polymers* 67(4): 530–535.

Khaliq, A., Chughtai, M. F. J., Mehmood, T., Ahsan, S., Liaqat, A., Nadeem, M., Ali, A. 2021. Chapter 3 - High-pressure processing; principle, applications, impact, and future prospective. In C. M. Galanakis (Ed.), *Sustainable Food Processing and Engineering Challenges* (pp. 75–108). London: Elsevier Academic Press.

Kim, H. S., Kim, B. Y., & Baik, M. Y. 2012. Application of ultra high pressure (UHP) in starch chemistry. *Critical Reviews in Food Science and Nutrition* 52(2): 123–141.

Kim, H. Y., Ye, S. J., & Baik, M. Y. 2023. Pressure moisture treatment (PMT) of starch, A new physical modification method. *Food Hydrocolloids* 134: 108051.

Knorr, D., Heinz, V., & Buckow, R. 2006. High pressure application for food biopolymers. *Biochimica et Biophysica Acta (BBA)-Proteins and Proteomics* 1764 (3): 619–631.

Larrea-Wachtendorff, D., Sousa, I., & Ferrari, G. 2021. Starch-based hydrogels produced by high-pressure processing (HPP): Effect of the starch source and processing time. *Food Engineering Reviews* 13(3): 622–633.

Leite, T. S., de Jesus, A. L. T., Schmiele, M., Tribst, A. A. L., & Cristianini, M. 2017. High pressure processing (HPP) of pea starch: Effect on the gelatinization properties. *LWT-Food Science and Technology* 76: 361–369.

Li, W., Bai, Y., Mousaa, S. A. S., Zhang, Q., & Shen, Q. 2012. Effect of high hydrostatic pressure on physicochemical and structural properties of rice starch. *Food and Bioprocess Technology* 5(6): 2233–2241.

Li, W., Gao, J., Saleh, A. S. M., Tian, X., Wang, P., Jiang, H., & Zhang, G. 2018. The modifications in physicochemical and functional properties of proso millet starch after ultra-high pressure (UHP) Process. *Starch-Stärke* 70(5–6): 1700235.

Li, W., Tian, X., Liu, L., Wang, P., Wu, G., Zheng, J., Zhang, G. 2015. High pressure induced gelatinization of red adzuki bean starch and its effects on starch physicochemical and structural properties. *Food Hydrocolloids* 45:132–139.

Liu, H., Fan, H., Cao, R., Blanchard, C., & Wang, M. 2016a. Physicochemical properties and in vitro digestibility of sorghum starch altered by high hydrostatic pressure. *International Journal of Biological Macromolecules* 92: 753–760.

Liu, H., Guo, X., Li, Y., Li, H., Fan, H., & Wang, M. 2016b. In vitro digestibility and changes in physicochemical and textural properties of Tartary buckwheat starch under high hydrostatic pressure. *Journal of Food Engineering* 189: 64–71.

Liu, H., Wang, L., Cao, R., Fan, H., & Wang, M. 2016c. In vitro digestibility and changes in physicochemical and structural properties of common buckwheat starch affected by high hydrostatic pressure. *Carbohydrate Polymers* 144: 1–8.

Muhr, A. H., & Blanshard, J. M. V. (1982). Effect of hydrostatic pressure on starch gelatinisation. *Carbohydrate Polymers* 2(1): 61–74.

Muhr, A. H., Wetton, R. E., & Blanshard, J. M. V. 1982. Effect of hydrostatic pressure on starch gelatinisation, as determined by DTA. *Carbohydrate Polymers* 2(2): 91–102.

Ratnayake, W. S., & Jackson, D. S. (2008). Chapter 5 - Starch gelatinization. *Advances in Food and Nutrition Research* (Volume 55; pp. 221–268). London: Elsevier Academic Press.

Stute, R., Heilbronn, Klingler, R. W., Boguslawski, S., Eshtiaghi, M. N., & Knorr, D. 1996. Effects of high pressures treatment on starches. *Starch-Stärke* 48(11–12): 399–408.

Thevelein, J. M., Van Assche, J. A., Heremans, K., & Gerlsma, S. Y. 1981. Gelatinisation temperature of starch, as influenced by high pressure. *Carbohydrate Research* 93(2): 304–307.

Vallons, K. J. R., & Arendt, E. K. 2009. Effects of high pressure and temperature on the structural and rheological properties of sorghum starch. *Innovative Food Science & EmergingTechnologies* 10(4): 449–456.

Wang, N., Dong, Y., Dai, Y., Zhang, H., Hou, H., Wang, W., Li, C. 2023a. Influences of high hydrostatic pressure on structures and properties of mung bean starch and quality of cationic starch. *Food Research International* 165: 112532.

Wang, N., Dong, Y., Zhang, H., Wang, B., Cao, J., Dai, Y., Zhang, Y. 2023b. Exploring the mechanism of high hydrostatic pressure on the chemical activity of starch based on its structure and properties changes. *Food Chemistry* 418: 136058.

Woldemariam, H. W., & Emire, S. A. 2019. High pressure processing of foods for microbial and mycotoxins control: Current trends and future prospects. *Cogent Food and Agriculture* 5(1): 1622184.

Xia, Q., Wang, L., Xu, C., Mei, J., & Li, Y. 2017. Effects of germination and high hydrostatic pressure processing on mineral elements, amino acids and antioxidants in vitro bioaccessibility, as well as starch digestibility in brown rice (*Oryza sativa* L.). *Food Chemistry* 214: 533–542.

Yashini, M., Khushbu, S., Madhurima, N., Sunil, C. K., Mahendran, R., & Venkatachalapathy, N. 2022. Thermal properties of different types of starch: A review. *Critical Reviews in Food Science and Nutrition* 1–24.

CHAPTER 4

Extrusion Treatment of Starch and Starchy Materials

Wen Ma, Huifang Cao, Enbo Xu, Shuohan Ma, and Haibo Pan

4.1 INTRODUCTION

Extrusion is a high-performance multi-functional process, and its application has been widely expanded in the continuous treatments of polymer chains like starch/cellulose/plastics since the 1960s (Figure 4.1). Basically, an extruder (either single-screw extruder, SSE, or twin-screw extruder, TSE) is used as a (bio)reactor with the capacities of adding feedstock/additives at moderate levels through the barrel module, with improved thermal and mechanical mixing, enhanced reaction efficiency, combining various operating functions, increasing devolatilization and offering direct (on-line)/downstream product shaping and structuring (Bouvier & Campanella, 2014; Brown, 1991; Govindasamy, Campanella, & Oates, 1997; Linko, Linko, & Olkku, 1983; Xu et al., 2015). Many reactions have been performed in extruder-reactors (Janssen, 2004; Xanthos, 1992), and their types (mainly (bio)degradation of bio-based virgin materials, functional modification, bulk polymerization and reactive solid–liquid extrusion-pressing) have prompted new case studies. For example,

 DOI: 10.1201/9781003493594-4

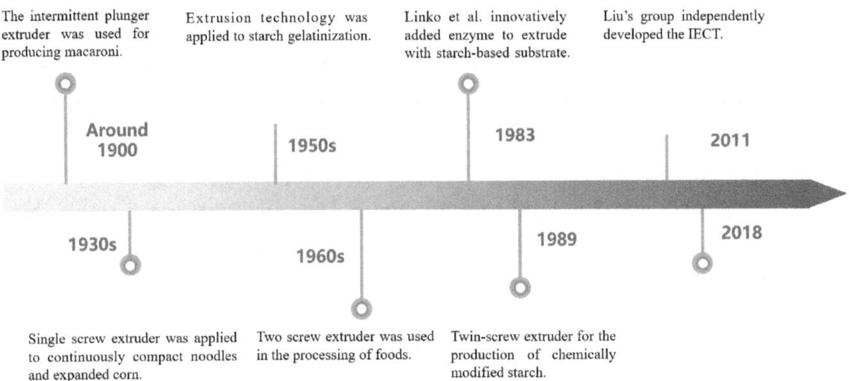

FIGURE 4.1 Timeline of important events in the field of starch-related extrusion.

depolymerization of extruded bio-based polymers is important for increasing the value of degraded products with/without the addition of compounds such as acids, alkalis, enzymes, etc. (Bouvier & Campanella, 2014; Duque et al., 2014; Xu, Wu, Jin, & Campanella, 2017a). Concerning polymer modifications, functionalization including grafting one or more monomers and introducing functional groups onto the polymer backbone, or modifying existing functional sites (Brown, 1991; Moad, 2011) are still research hotspots. Bulk polymerization, which synthesizes new high-order polymers from monomers or from mixtures of different original polymers, often leads to macromolecular products with higher molecular weight (M_w) in little or no solvent environments using extrusion (Bouvier & Campanella, 2014; Gandini & Lacerda, 2015).

Starch and starchy plants are often investigated as main extrusion feedstocks. Extrusion was first utilized in the production of macaroni, noodles and cereal instants in starch-based foods, and then it was gradually used for the creation and/or reuse of modified starches. As early as the 1950s, extrusion technology was applied to starch gelatinization, which could be used directly or indirectly for further process. In 1983, Linko et al. (1983) innovatively added an enzyme to extrude with starch-based substrate and demonstrated that thermostable α-amylase could accelerate the liquefaction of starch (i.e. enzymatic extrusion or bioextrusion). In non-food applications, extruded starch has been widely used in recent years for the preparation of biodegradable materials (e.g. starch-based films, foams, straws etc.). In addition, Xu, Wu, Long, Jiao, and Jin (2018b) used bioextrusion to prepare novel porous starch (see Section 4.4), which could be used as adsorption materials. All in all, the advantages of reactive extrusion for different materials have been thoroughly reviewed by many books and papers (Bouvier & Campanella, 2014; Janssen, 2004; Vergnes & Berzin, 2004, 2006), which are not the main focus of this book. Thus, the important message to be delivered in the present chapter is to describe the structural change mechanisms of starch during the melting and mixing in the extrusion process, and how to expand the utilization of green (bio) extrusion (and chemical extrusion has not been summarized here) in the production of starch-related food and non-food applications.

4.2 PRINCIPLE OF COMPLEX THERMOMECHANICAL ACTION IN EXTRUSION

4.2.1 Reaction Engineering Fundamentals of Extruder

Both single- and twin-screw technologies are well-recognized and used in the starch industries. Recently, the TSE technology especially incorporating intermeshing, co-rotating design has been preferred for biopolymers like starch in comparison with the SSE technology due to several reasons including: (1) TSE provides a high degree of micromixing between components at the molecular and mesoscale levels and shows better-mixing capability than SSE which exhibits heterogeneous mixing of reaction components due to the different trajectories and velocities in the screw channel. (2) TSE exhibits favorable heat transfer for high-viscosity extrudates in the extrusion operation. (3) Materials are subjected to a less residence time distribution (RTD), where starve feeding is applied (see below) (Berzin, Tara, Tighzert, & Vergnes, 2010; Bouvier & Campanella, 2014; Janssen, 2004; Moad, 2011). Furthermore, limited screw wear does not affect the performance of TSE while it significantly decreases the productivity, consistency and flexibility of SSE.

TSE technology combined with ancillary equipment usually comprises six main parts, i.e., preconditioner, power drive (motor, gearbox, etc.), central operating cabinet, screw-barrel assembly, die assembly and downstream production equipment (Bouvier & Campanella, 2014). The screw-barrel assembly is the TSE core zone where the physical and chemical reactions take place. The transformation of starch materials from a solid/particulate state to a semifluid state in an extrusion process is schematically illustrated in Figure 4.2. A generic extrusion process consists of a conveying or named feed/transport section, a mixing section (including compression, shearing and pressure build-up sections) and a metering section (where in addition to shear heating or cooling can be applied) which go from the mixing zone to the end of barrel connected with die

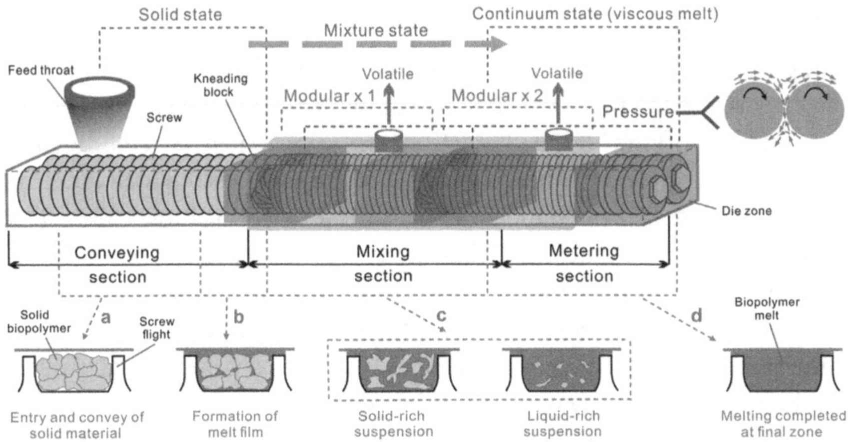

FIGURE 4.2 Chemical reaction sections and melting mechanism of biopolymers (especially starch) in an intermeshing co-rotating twin-screw extruder.

zone. Modular zones 1–2 are repeated zones, where the interpenetration and interaction between the different (macro)molecular phases with/without a degassing operation take place during a time–shear–temperature treatment. It is noteworthy that there is a significant distinction between thermal-processing machines and screw extruders where high pressure coupled with mechanical stress is built up (Figure 4.2).

During extrusion processing, the starch particle is firstly subjected to thermomechanical action and pressure build-up near the first kneading block with a compression action. Those mainly mechanical forces gradually convert the solid phase into a mixture of particles and a continuum of viscous melt. A hypothesis on starch melting mechanism based on biopolymer (Todd, 1993; Vergnes, Souveton, Delacour, & Ainser, 2001) is elucidated and schematically illustrated in Figure 4.2. Starch in the TSE process experiences rapid melting over rather short lengths (shown as stage "b"). Part of the molten starch sticks to the barrel wall and forms a thin film at relatively low pressures. Then mixing with increasing of the semifluid phase and reacting of material ingredients dispersed in that phase occurs (stage "c") in the mixing section, and the thermophysical action includes compression, shearing and heating. At the final stage "d", the thermophysical action is performed effectively as the interpenetration of the screws and the extruded starch material reaches a stage that is suitable to favor biochemical reactions.

4.2.2 Theoretical and Empirical Modeling

Due to the complex conversion discussed above, it is hard to consider all the aspects involved in the extrusion process for achieving the reaction objectives and identifying all the interactions between system variables such as heat transfer, residence time, phase state of the melt viscous state, mixing performance, degree of (de)polymerization, etc. (Berzin, Tara, Tighzert, & Vergnes, 2010; Farahanchi & Sobkowicz, 2017; Vergnes & Berzin, 2006). And that is one of the reasons for the lack of relevant extrusion kinetics involved in those processes. The approach used to overcome most of those limitations has been the use of "black box" extrusion systems that use and combine the extensive experimental data existing in the literatures.

Numerous kinetic models (some proven to be not accurate) fitting the extrusion performance and behavior are presented as conceptual flow models that provide a benchmark of ideal systems with a number of assumptions that greatly simplify the simulation of extrusion processes (Xu et al., 2017b). The basic approach for modeling extruder-reactors has been built on well-established models that describe plug flow reactor (PFR) and continuous stirred tank reactor (CSTR). However, these models consider an oversimplification of the melting flow conditions with rapid and low-accuracy computations for process control applications (Bouvier & Campanella, 2014; Vergnes & Berzin, 2006). In that sense, they could not be used to meet the need to scale-up processes. Vergnes and Berzin (2006) attempted to summarize a method that shows a numerical model with high accuracy in a wide range of processing conditions to extrusion. Based on the challengeable information of complex geometry, reaction kinetics and evolutionary, they focused on the models based on continuum mechanics as reported. Other authors also combined mathematical models with experimental extrusion results to estimate the evolution of the chemical reactions (Farahanchi & Sobkowicz, 2017; Grimard, Dewasme, Thiry, Krier, Evrard, & Wouwer, 2016). Suitable kinetic models associated with extrusion processes require a precise estimation of the reaction times which are normally incorporated using the RTD concept.

RTD provides the appropriate age distribution of the material inside the extruder. It is commonly measured by following the concentration evolution of a tracer introduced in the extrusion process. Knowledge of the process of RTD shows great benefits to describe the extrusion flow and mixing behavior of starch materials. It also allows the estimation of mean residence time (MRT, $\bar{t}$) of the physical–chemical reactions occurring in the process (Bouvier & Campanella, 2014). As discussed, determination of RTD is performed by following the evolution of the tracer concentration with time, *C(t)*, from which several RTD functions such as the exit age distribution *E(t)* (s^{-1}) and its indefinite integral *F(t)*, $\bar{t}$ and the standard deviation σ^2, as well as its normalized time (dimensionless parameter) expressed as θ ($= t/\bar{t}$) often used to regularly compare various RTD curves, can be calculated as follows (Chuang & Yeh, 2004; Levenspiel, 1972):

$$E(t) = \frac{C(t)}{\int_0^\infty C(t)dt} \cong \frac{C(t_i)}{\sum_0^\infty C(t_i)\Delta t_i} \tag{4.1}$$

$$F(t) = \int_0^t E(t)dt \cong \frac{\sum_0^t C(t_i)\Delta t_i}{\sum_0^\infty C(t_i)\Delta t_i} \tag{4.2}$$

$$\bar{t} = \int_0^\infty tE(t)dt = \frac{\sum t_i C_i \Delta t_i}{\sum C_i \Delta t_i} \tag{4.3}$$

$$\sigma^2 = \int_0^\infty (t-\bar{t})^2 E(t)dt = \frac{\sum (t_i-\bar{t})^2 C_i \Delta t_i}{\sum C_i \Delta t_i} = \frac{\sum t_i^2 C_i \Delta t_i}{\sum C_i \Delta t_i} - \bar{t}^2 \tag{4.4}$$

$$E(\theta) = \bar{t} \times E(t) \tag{4.5}$$

$$F(\theta) = F(t) \tag{4.6}$$

Furthermore, the mixing action of simulated extrusion reactor could be estimated by semi-empirical models related to the axial dispersion of the reactive compounds using a model describing the transport of each reaction component *i* through the extruder-reactor with time-dependent concentration $C_i(t)$ and an associated empirical lumped dispersion coefficient D_a. The following equations are used to estimate the axial dispersion of each *i* and the empirical dispersion coefficient D_a estimated with dimensionless parameters associated with mass transfer such as the Peclet number (*Pe*) (Froment, Bischoff, & Wilde, 1990):

$$\bar{u}\frac{\partial C_i(t)}{\partial z}=\frac{\partial}{\partial z}\left(D_a\frac{\partial C_i(t)}{\partial z}\right)+r_i(C_i,C_j,\ldots\ldots C_n) \tag{4.7}$$

$$D_a=\frac{\bar{u}\,L}{Pe} \tag{4.8}$$

where $\bar{u}$ is the z-component of the mean velocity through the extrusion reactor, L is the characteristic length of the extrusion reactor and r_i $(C_i, C_j \ldots C_n)$ is the biochemical reaction associated to the component i. Pe and D_a are the opposite counterparts to quantify the degree of axial dispersion, i.e., $Pe \to \infty$ (or $D_a \to 0$) means an ideal PFR while $Pe \to 0$ (or $D_a \to \infty$) means a perfect mixing reactor (Bouvier & Campanella, 2014; Chuang & Yeh, 2004), whereas extruders are characterized by parameters with values that lie between those extreme and ideal conditions. Herein, the conversion of starch in an extrusion reactor could change from that achieved in PFR to that achieved in a perfectly mixing reactor when D_a increases to large values. De Ruyck (1997) proposed that the extent of axial mixing during extrusion can also be estimated by the dimensionless number (i.e. the number of CSTR expressed as N) obtained from parameters derived from the RTD function as

$$\sigma_\theta{}^2=\frac{\sigma^2}{\bar{t}^2}=\frac{1}{N}=\frac{2}{Pe}-\frac{2}{Pe^2}(1-e^{-Pe})=2\frac{D_a}{\bar{u}\,L}-2(\frac{D_a}{\bar{u}\,L})^2(1-e-\frac{\bar{u}\,L}{D_a}) \tag{4.9}$$

In considering actual micromixing and the rheology of the extruded starch material, its flow pattern during extrusion is a non-Newtonian reactive fluid involving reaction kinetics lying between those occurring in plug flow and perfect mixing reactors (Altomare & Ghossi, 1986; Ficarella, Milanese, & Laforgia, 2006; Kumar, Ganjyal, Jones, & Hanna, 2008; Lee & Mccarthy, 1996). To estimate the RTD directly, many mathematical models to fit those curves were developed as follows (Table 4.1): (a) the laminar flow in a pipe (LFP) model (Wolf & White, 1976); (b) the Logistic model (Yu, Meng, Ramaswamy, & Boye, 2014); (c) the CSTR-in-series model (Apruzzese, Pato, Balke, & Diosady, 2003; De Ruyck, 1997); (d) the Wolf-Resnick model (Wolf & Resnick, 1963); (e) the Wolf-White model (Wolf & White, 1976); (f) the Yeh-Jaw complete model (Yeh & Jaw, 1999); and (g) the Yeh-Jaw simplified model (Yeh & Jaw, 1999). For example, Xu et al. (2017b) investigated the RTD of rice during extrusion simultaneously under enzymatic degradation and compared the effectiveness of different flow models in fitting the RTD data.

TABLE 4.1 Common Mathematical Models Fitting the Mixing Curves of Extrusion

Models	$F(\theta)=\int E(\theta)$	Parameters	References
LFP	$F(\theta)=1-\left(\frac{1}{2\theta}\right)^2$; $\theta \geq 0.5$	/	Wolf and White (1976)
Logistic	$F(\theta)=\frac{U}{(1+e^{-B(\theta-M)})}$; $\theta \geq 0$	U=the upper limit of tracer concentration B=the tracer accumulation rate M=the internal age of half-concentration	Yu, Meng, Ramaswamy, and Boye (2014)
CSTR-in-series	$F(\theta)=1-e^{-N\theta}\left[1+N\theta+\frac{1}{2!}(N\theta)^2+\cdots+\frac{1}{(N-1)!}(N\theta)^{N-1}\right]$; $\theta \geq 0$	N=the number of CSTR	Apruzzese, Pato, Balke, and Diosady (2003); De Ruyck (1997)
Wolf-Resnick	$F(\theta)=0$; $0<\theta<\varepsilon/\overline{t}$ $F(\theta)=1-e^{-\eta(\theta-\varepsilon/\overline{t})}$; $\theta \geq \varepsilon/\overline{t}$	η=the coefficient of exponent ε=the system phase shift $\overline{t}$=the MRT	Wolf and Resnick (1963)
Wolf-White	$F(\theta)=0$; $0<\theta<P$ $F(\theta)=1-e^{-(\frac{1}{1-P})(\theta-P)}$; $\theta \geq P$	P=the fraction of PFR	Wolf and White (1976)
Yeh-Jaw complete	$F(\theta)=0$; $0<\theta<P$ $F(\theta)=1-\frac{dm_1m_2-bm_2}{b(m_1-m_2)}e^{-m_1(\frac{\theta-P}{1-P})}+\frac{dm_1m_2-bm_1}{b(m_1-m_2)}e^{-m_2(\frac{\theta-P}{1-P})}$; $\theta \geq P$ $m_1,m_2=\frac{(b+d)\pm\sqrt{(b+d)^2-4bd(1-d)}}{2d(1-d)}$	b=the fraction of effluent cross-flowing between CSTR and dead volume d=the fraction of dead volume in CSTR P=the fraction of PFR	Yeh and Jaw (1999)
Yeh-Jaw simplified	$F(\theta)=0$; $0<\theta<P$ $F(\theta)=1-e^{-\frac{(\theta-P)}{(1-P)(1-d)}}$; $\theta \geq P$	P=the fraction of PFR d=the fraction of dead volume in CSTR	Yeh and Jaw (1999)

4.3 STRUCTURAL CHANGE OF (BIO) EXTRUSION-TREATED STARCH

Extrusion (and bioextrusion) can lead to the structural changes (e.g. gelatinization, degradation) of starch with a partial or total destruction of its hierarchical structure, thereby changing the properties of starch and making it more widely used with modification. The irreversible and physical transition process of starch can be positively accelerated by mechanical actions of friction, compacting and shearing in the extrusion process. The main objective of this chapter is to elaborate on the effects produced by thermomechanical action on the different structures of starch as (bio) extrusion-treated starch (ETS/bioETS).

4.3.1 Starch Granule Morphology

Extrusion at high temperature, high pressure and high shear breaks the starch granules and destroys their structure. The ETS form an irregular hole-like structure (mainly puffed) due to the sudden drop in pressure and evaporation of water during the extrusion process. Almost all starch granules lose their integrity after extrusion (Huang, Liu, Ma, Mai, & Li, 2022; Pérez & Bertoft, 2010). Natural corn starch granules are polygonal with relatively smooth edges, but after extrusion, the starch structure was completely destroyed and the granules were severely damaged with the un-smooth surface, irregular fragments and cracks (Yan, Wu, Li, Yin, Ren, & Tao, 2019). Wu, Jiao, Xu, Chen, and Jin (2020) also observed numerous small pores on the surface of bioETS.

4.3.2 Starch Ordered Structure

Extrusion involves both thermal and mechanical energies, and causes damage to both the short-range ordered structure (double helices) and the long-range ordered structure (crystals) of starch. For the double-helical structure, ETS rapidly absorbs water and swells, leading to the dissociation of branched starch double helices (Liu, Halley, & Gilbert, 2010). On the other hand, ETS chains can also recombine into the double helices during the cooling process (i.e. retrogradation), which means a transition between ordered and disordered structures. The crystalline region of ETS is to some extent destroyed with the amorphous region increased under the action of heat and shear force. For example, Zhang et al. (2023) found that the crystallinity of starch in Tartary buckwheat seeds decreased significantly after extrusion. Compared with ordinary starch or waxy starch, high-amylose starch can completely or partially maintain its crystallinity structure after extrusion, which may be related to its high gelatinization temperature (Shrestha, Ng, Lopez-Rubio, Blazek, Gilbert, & Gidley, 2010).

Extrusion also changes the crystal form of starch. In the process of extrusion, the four crystal forms of A-, B-, C- and V-types may be converted to each other due to the gelatinization and recrystallization of starch, as well as the addition of objects such as aliphatic chains and phenols (Ali, Singh, & Sharma, 2020; Yang et al., 2016). Liu et al. (2017) extruded rice starch by an SSE, and the crystal of rice starch changed from A-type to V-type, which may be due to the amylose–lipid complex formed during grain starch extrusion. According to the findings of Wang, Wu, Zhang, Kan, and Zheng (2022), though rice starch is A-type crystal form, its crystal structure was entirely shattered with a new structure of A- and V-type that occurred during extrusion. Table 4.2 shows the effects of different extrusion conditions on the crystallinity and crystal form of starches.

4.3.3 Starch Molecular Structure

The degree of degradation of ETS depends on the type of starch and the change of mechanical energy during extrusion (Li, Hasjim, Xie, Halley, & Gilbert, 2014; Zhang, Dhital, Flanagan, Luckman, Halley, & Gidley, 2015). The high temperature of extrusion causes the starch to absorb heat and gelatinize, and the molecular chains are scattered while the mechanical shear force attacks the hydrogen bonds of starch. Due to the large molecular weight and high branching degree of amylopectin, its α-1,6-glycosidic bonds are liable to break under the action of shearing force. Amylopectin is more likely to be degraded during extrusion compared to amylose which is hardly degraded regardless of extrusion processing conditions (Li, Hasjim, Xie, Halley, & Gilbert, 2014; Liu, Chen, Wu, Luo, Chen, Liu, et al., 2019). The higher the content of amylose is, the smaller the change of size distribution is, showing that amylopectin is more easily degraded than amylose during extrusion. However, the amylopectin molecule does not degrade infinitely by extrusion, which is degraded continuously to approach the maximum stable size (Liu, Halley, & Gilbert, 2010). The degradation pattern of ETS molecules is shown in Figure 4.3, the degradation happens primarily near the branching point and the molecular size distribution of amylopectin is identical (Liuet al., 2010). Although the size distribution of amylopectin changed significantly, the chain-length distribution of amylopectin as well as amylose (though broken with the α-1,4 glycosidic bonds by extrusion sometimes) did not change dramatically (Liu et al., 2019). Li et al. (2021) found that when jackfruit seed starch was modified by improved extrusion

TABLE 4.2 Changes of the Crystallinity of Starch after Extrusion

Raw Material	Extruder Model	Water Content (%)	Screw Speed (r/min)	Sleeve Temperature (°C)	Crystallinity and Crystal Form	References
Corn starch	Twin-screw extruder (Clextral, BC 21)	14, 18	100	100	Decrease, change from A-type to V-type	Ali, Singh, & Sharma (2019)
Potato starch		22	100	100	Decrease, change from B-type to weak B-type	Ali, Singh, & Sharma (2019)
Potato starch		18	100	160	Decrease, change from A-type to C-type	Ali, Singh, & Sharma (2020)
Rice starch	Self-made single screw	30, 40, 50, 60, 70	37.5	50, 65, 85, 100, 95	Decrease, change from A-type to V-type	Liu et al. (2017)
Potato starch	Twin-screw extruder (Eurolab 16, Thermo Fisher Scientific Co., Ltd.)	40	50	60, 70, 70, 80, 80, 110	Decrease and the crystal form disappears (amorphous)	Sun et al. (2021)
Bamboo shoot dietary fiber and rice starch	Twin-screw extruder (SLJ-50A, Jinan Huinuo Machinery Technology Co., Ltd.)	50	40	90, 110, 80	Reduced from 42.43% to 3.52%, change from A-type to A+V-type	Wang, Wu, Zhang, Kan, & Zheng (2022)
Corn starch and lecithin	Coax ial twin screw	30	160	60, 75, 90, 105, 120	Decrease, change from A-type to A+V-type	Yang et al. (2016)
Tartary buckwheat flour	Twin-screw extruder (Brabender KETSE 20/40, Duisburg)	30	100	40, 75, 110, 160, 95	Reduced from 43.81% to 12.95%, change from A-type to V-type	Liu et al. (2023)

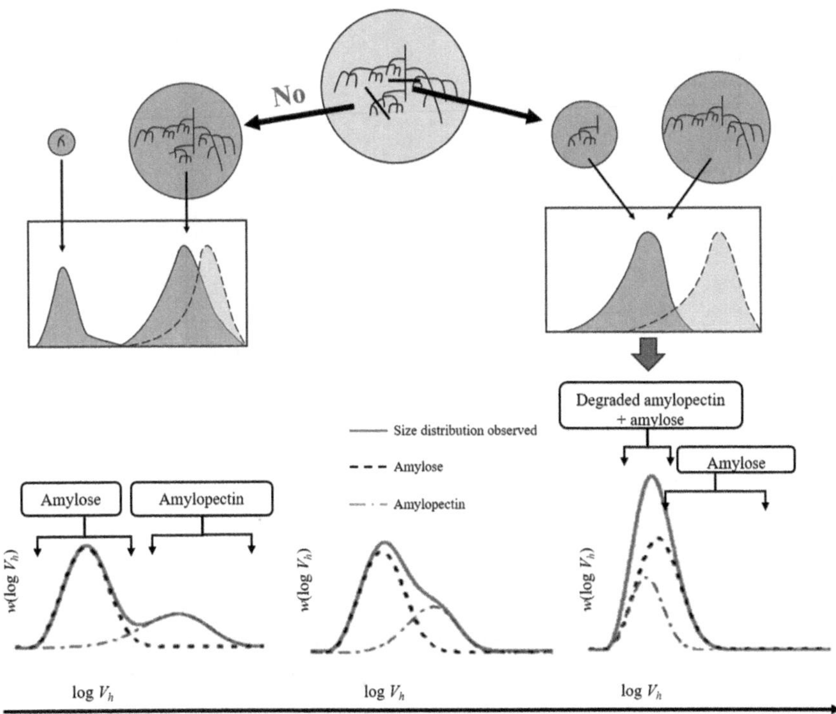

FIGURE 4.3 Sketch of degradation pattern of starch molecules during extrusion. Reproduction from Liu, Halley, and Gilbert (2010), with permission.

cooking technology, the α-1,4-glycosidic bond in amylopectin was broken, resulting in the decrease in the content of long chains and the increase in short and medium chains. According to the findings of Brümmer et al. (2002), considerable breakage of α-1,4-glycosidic bond would occur only when product temperatures exceeded 180°C.

Furthermore, the degradation of starch was greater under bioextrusion in the presence of amylase (Xu et al., 2016). Natural starch has high structure strength (especially in tissue matrix) and crystallinity, and it is difficult for the enzyme to attack deeply. Extrusion effectively destroys the natural and orderly tissue structure of the starchy raw material, and the contact surface area between the enzyme and the substrate increases simultaneously, showing high enzyme activity for bioETS (Figure 4.4). In addition, short chains (DP < 12) have relatively strong diffusivity in solution, and they may to some extent weaken the enzyme efficiency of amylase, which can be avoided by the low water environment provided by bioextrusion

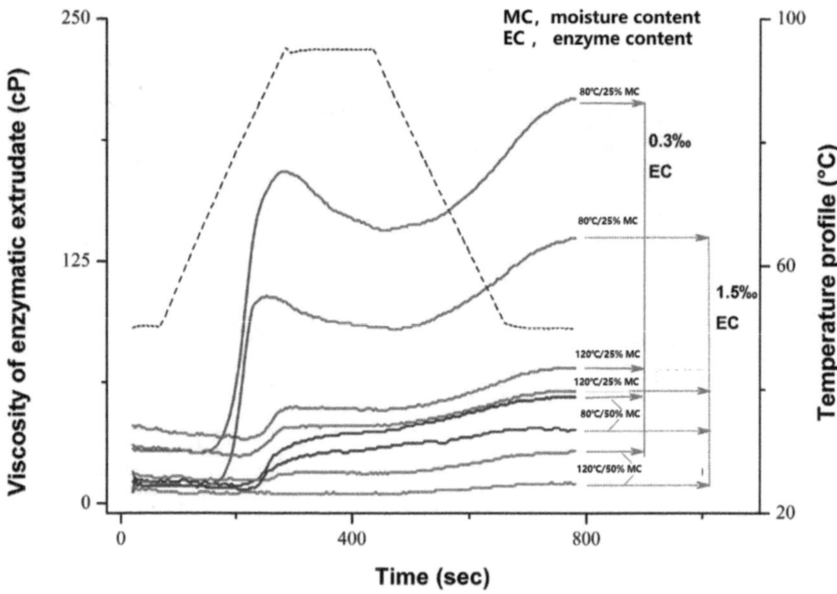

FIGURE 4.4 Viscosity curves and total phenolic contents for enzymatic extruded rice with different enzyme concentrations, moisture contents and treated temperatures. Reproduction from Xu et al. (2016), with permission.

(or called enzymatic extrusion) technology (Wang, Zhang, Chen, & Zhong, 2020; Wang, Zhang, He, Zhang, Zhou, Wang, et al., 2019).

4.4 ETS/BIOETS APPLICATIONS

4.4.1 Saccharification and Fermentation

Grains (rice, corn, barley, wheat, etc.) have high starch content, and they can be extruded and deeply processed into starch sugar, syrup, alcohol and other fermented products (Offiah, Kontogiorgos, & Falade, 2018). However, due to the dense cell wall and complete structure of macromolecules, there are some problems in their processes such as insufficient contact with enzymes or microorganisms. Extrusion/bioextrusion can degrade starch-based materials as well as protein and fat in them to a certain extent (Xu, Campanella, Ye, Jin, Liu, & BeMiller, 2020a). During extrusion, starch absorbs water and swells, and its crystal and double helix structures are destroyed and small molecular substances such as oligosaccharides are generated (Ye et al., 2018). Therefore, ETS/bioETS meets the production needs of saccharification and fermentation industries, especially in the fields of sugar refining and brewing. Kuo, Shieh, Huang, Wang, and Huang (2019) reported that the native crystalline structure of brown rice was disrupted after extrusion, and the rate of enzymatic hydrolysis was improved with high saccharification efficiency and post-fermentation rice wine alcohol yield. ETS/bioETS granules were found to break and agglomerate, with many hydrolysis pores on their surface, and the molecular weight of these extrusion-puffed barley decreased greatly (Zhao, Jiao, Yang, Liu, Wu, & Jin, 2022). Chien, Tsai, Wang, Dong, Huang, and Kuo (2022) pretreated brown rice, corn and buckwheat grains with high starch content by extrusion for the feasibility of manufacturing high-maltose syrup. Furthermore, salts such as Ca, Mg and other metal ions have been found to show a significant influence on the formation of bioETS samples recently, and their fermentation efficiencies can be greatly changed by these salts with different anions (Xu et al., 2020b, 2020c; Xu, Wu, Jiao, & Jin, 2018a; Xu et al., 2017a) (Figure 4.5).

In addition, fermentation of ETS/bioETS is widely used in the chemical industry, especially for some special purposes including high-value-added products. Chen, Don, and Yen (2006) found that corn starch can contain up to 34.1% reducing sugars when extruded under the appropriate conditions, which has almost zero reducing sugars in natural starch. The addition of ETS to the medium for the cultivation of *Haloferax mediterranei* showed the highest growth rate, shorter delay time and higher cell concentration compared with other set groups, confirming that maize starch treated with appropriate extrusion conditions is a better carbon source for the growth of *Haloferax mediterranei* cells and their production of polyhydroxyalkanoate. Another case is that Fasheun, da Silva, Teixeira, and Ferreira-Leitão (2023) discovered that employing extruded cassava starch for dark fermentation hydrogen generation reduced the adaptive fermentation period by 76% and raised the maximum H_2 yield by 43% when compared to the untreated counterpart.

4.4.2 Functional Starch-Based Food

The structure and properties of starch can be changed complicatedly during extrusion, relating to the digestibility, texture, cooking quality and other functional properties for the needs of different people. Gulzar, Hussain, Naseer, and Naik (2021) used extrusion technology to modify rice flour by adjusting feed moisture, barrel temperature and screw speed and found that the melting temperature of the crystallization zone of ETS

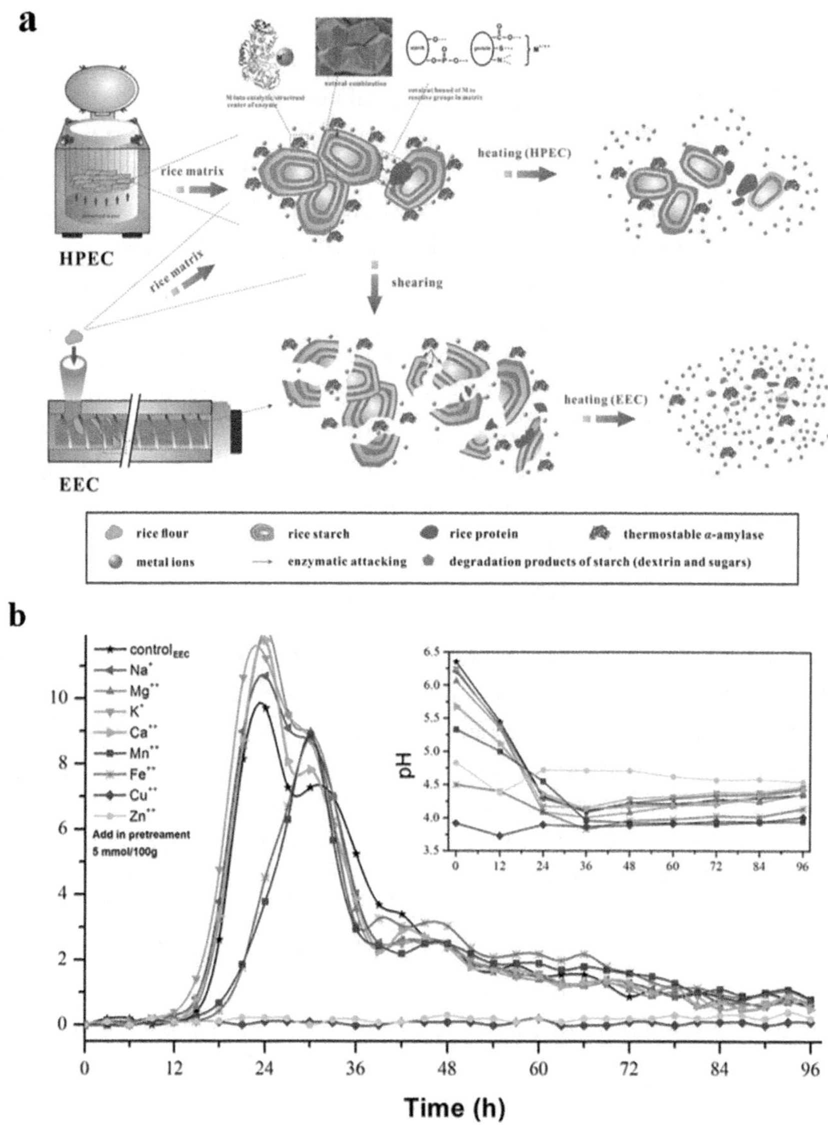

FIGURE 4.5 (a) Comparison of EEC (enzymatic extrusion/bioextrusion) and HPEC (high-pressure enzymatic cooking) for starch degradation. (b) Effects of various metal ions (5 mmol/g) on alcoholic fermentation rate expressed as CO_2 loss by EEC pretreatment of rice. Reproduction from Xu et al. (2018a), with permission.

was higher than the native one. Also, slow digestible starch and resistant starch (RS, increased from 3.0% to 6.2%) of rice ETS increased significantly. Tang et al. (2022) produced a rearranged supramolecular starch structure by changing the ratio of amylose to amylopectin and adding pullulanase during extrusion, which has a great digesting resistance (Figure 4.6). These ETS/bioETS can be used to develop functional foods, especially for patients with diabetes. Besides, Li et al. (2018) created a porous structure for high-strength extruded noodles (HENs) by adding 0.05%–0.10% of thermostable α-amylase, improving the rehydration characteristics of HENs. Wang et al. (2019) also found that extrusion degraded the starch molecules in buckwheat flour, resulting in a shortening of the chain length and further changing its dynamic viscosity, which can be used as a binder or diluent to affect the gluten network form.

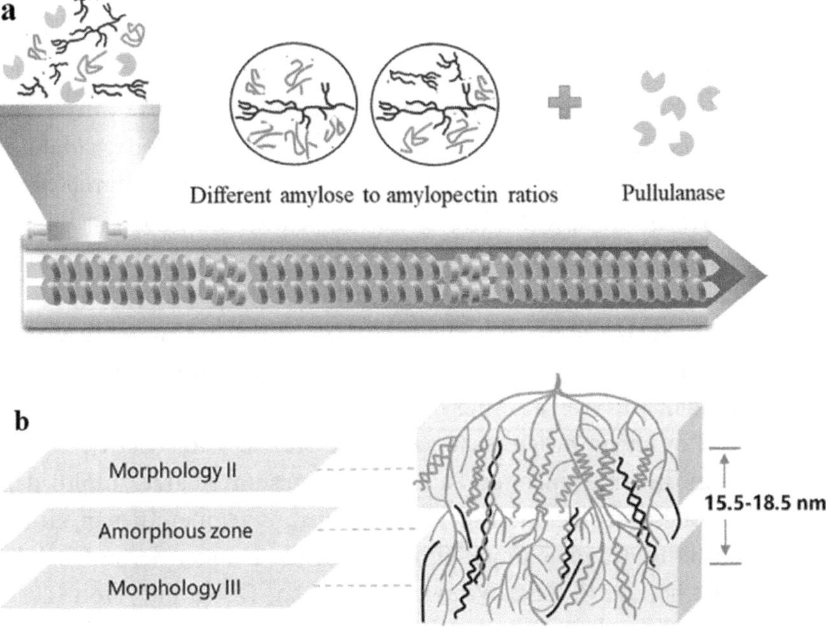

FIGURE 4.6 Schematic diagram for a high-solid biosynthesis strategy to prepare RS with different structural units or micro-domains. (a) Extruder-bioreactor of high-solid dynamic pullulanase catalysis. (b) Proposed multiple resistant supramolecular structure. Reproduction from Tang et al. (2022), with permission.

Simultaneously, starch can be employed as a carrier in the extrusion process to combine with polyphenols and other chemicals for the production of functional foods. Qin et al. (2023) found that bioextrusion with added mesophilic α-amylase effectively retained phenolics in brown rice compared to steaming and traditional extrusion. Ma et al. (2022) increased the retention of exogenously added ferulic acid by introducing thermostable α-amylase during extrusion, and the starch–ferulic acid complexes obtained had good in vitro release properties. Zheng, Tian, Ogawa, Yin, Xu, Chen, et al. (2021) significantly reduced the digestibility of rice by co-extruding proanthocyanidins from Chinese bayberry leaves. In addition to the polyphenols mentioned above, fatty acids, vitamins, etc. can also be added to starch as a formula for extrusion to prepare functional foods with low glycemic value, antioxidant and other properties (Dircio-Morales et al., 2022; Torres-Aguilar, Hayes, Yepez, Martinez, & Hamaker, 2023).

4.4.3 Biodegradable Materials

Starch has become a research hotspot for degradable materials by extrusion in recent years, due to its wide source, low cost, as well as biodegradability and processability (Versino, Lopez, Garcia, & Zaritzky, 2016). Starch-based films are one of the applications, but their brittleness, poor mechanical and barrier qualities are the drawbacks. Edible starch films with optimal mechanical and barrier properties could be created by co-extrusion of high amylose corn starch with the plasticizer glycerol (Calderón-Castro et al., 2018). Its use in the preservation of mangoes significantly reduced the loss of quality as well as other qualities. The plasticized starch can also be achieved by using other plasticizers like stearic acid, triethyl citrate, glyceryl monostearate, diisodecyl adipate, magnesium stearate, malto-dextrins, sorbitol, isosorbide, malic acid, low transition temperature mixtures, etc. (Montilla-Buitrago, Gómez-López, Solanilla-Duque, Serna-Cock, & Villada-Castillo, 2021). Zhai, Wang, Zhang, Dai, Dong, and Hou (2020) prepared starch–poly(butylene adipate-co-terephthalate) (PBAT) composite membranes by extrusion blow molding, and its tensile strength and elongation at break were significantly higher than those of pure starch membranes. Also, starch-based films with antimicrobial and barrier properties that can be used in bioactive and intelligent packaging can also be produced by extrusion with the addition of antimicrobial agents and hydrophobic active ingredients (Collazo-Bigliardi, Ortega-Toro, & Chiralt Boix, 2018; Cui, Ji, Wang, Xiong, & Sun, 2021) like cellulose fibers (Figure 4.7).

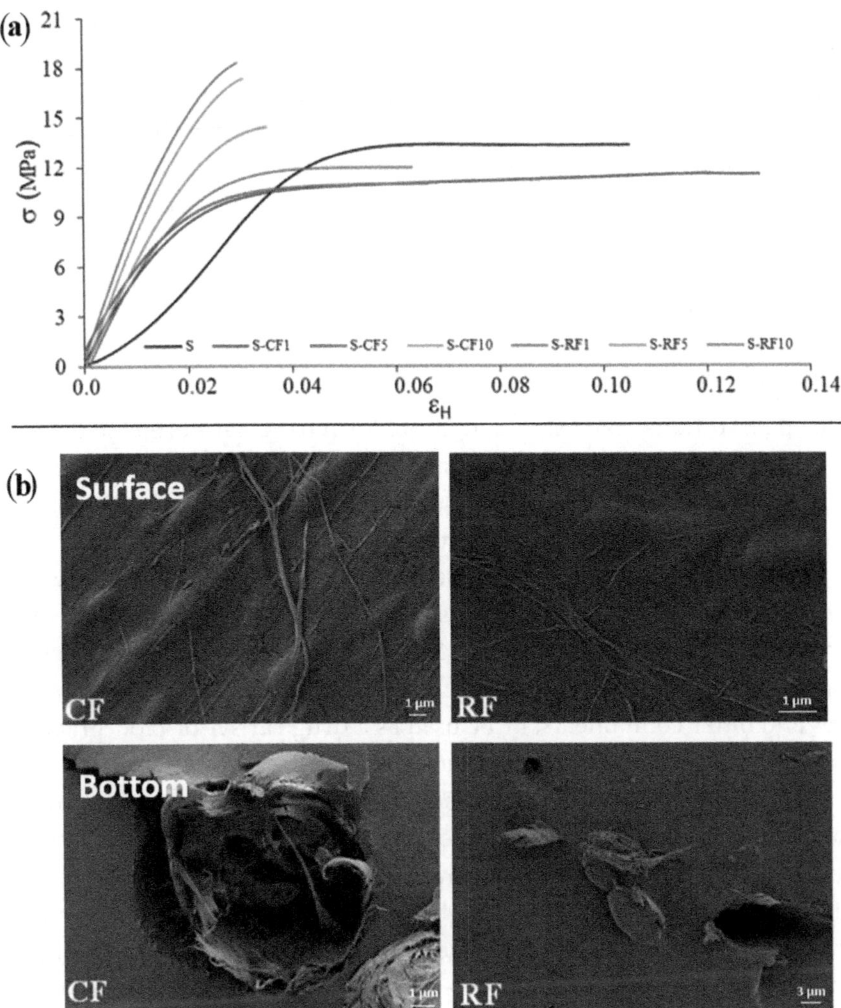

FIGURE 4.7 (a) Typical stress-Hencky strain curves for starch films with 1–10 wt% of cellulose fibers from coffee (CF) and rice (RF) husk. (b) Microscopies of fibers into the starch matrices, on the surface (top) and inside the film (bottom). Reproduction from Collazo-Bigliardi et al. (2018), with permission.

The production of starch-based foam is another application of ETS. According to Duan et al. (2022), starch-based foam was prepared by extruding hydroxypropyl corn starch using water as a foaming agent. During the extrusion process, the plasticizers of glycerin, poly(vinyl alcohol) and starch were cross-linked to form a three-dimensional network structure. Starch crystals and cellulose crystals were added as nucleating agents to promote uniform cells and improve compressive strength and elastic modulus, and using this low water absorption, high elasticity was obtained for loose filler, insulation, shock-proof boxes, etc. Georges, Lacoste, and Damien (2018) obtained a stabilized extruded foam by premixing starch with gluten, glycerol, yeast and sodium bicarbonate. Sodium bicarbonate increases the nucleation of the cells and stabilizes foam expansion, gluten improves foam elasticity and their proportions can be adjusted to produce foam with optimal expansion ratios and rheological properties.

4.4.4 Adsorption Materials

During traditional extrusion, starch can form irregular porous structures due to the sudden drop in pressure and evaporation of water. These porous structures will show different characteristics depending on the source and variety of raw starch and the extrusion parameters. In the case of high water content, there are fewer pores, but the pore wall is thicker, while high-temperature extrusion increases the number of pores and thins the pore wall (Lazou & Krokida, 2010). Starch with a suitable porous structure can be used to encapsulate flavor agents, bind to food to absorb impurities and bitter components, or be used as a drug carrier or biodegradable adsorbent, etc. (Zeller, Saleeb, & Ludescher, 1998).

Furthermore, Xu, Wu, Long, Jiao, and Jin (2018b) exogenously added zinc ions into natural corn starch to make zinc-modified starch, and then mixed it with (thermostable α-amylase)-magnesium complex for extrusion. Zinc ions were used to strengthen the backbone of starch and magnesium ions were used to stabilize the enzyme, and the starch with a macroporous structure was prepared, which solved the problem of starch quick liquefaction and difficulty to build a supportive structure in the process of biological extrusion (Figure 4.8). The water and oil absorption of the obtained porous starch were improved to different degrees. Xu, Wu, Ding, Ye, Jin, and Liu (2019) further co-extruded zerovalent iron particles and Fe_3O_4 nanoparticles with zinc-modified starch to create magnetic and porous biomaterials that, when used as adsorbents, removed dye better than pure starch. It may be used in biological and environmental applications. In addition, Wei, Wu, Ren, Yu, and Sun (2024) proposed a novel ion-esterification cooperative crosslinking-extrusion combined modification method. The obtained starch gels have smaller pore size distribution and denser honeycomb porous structure.

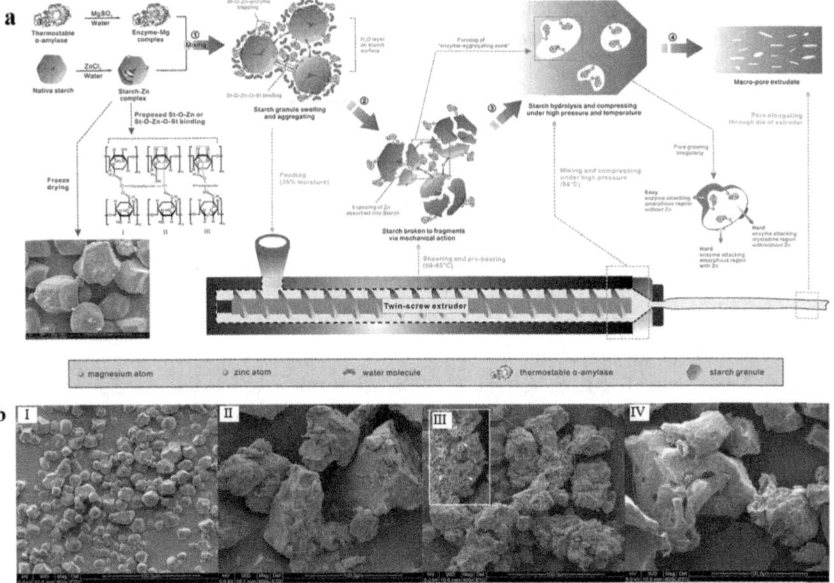

FIGURE 4.8 (a) Schematic illustration of Zn/Mg ions assisting a bioextrusion process (using thermostable α-amylase) to create macropores in starch. (b) SEM images of (I) native, (II) bioextruded, (III) Mg-bioextruded and (IV) Zn/Mg-bioextruded corn starches. Reproduction from Xu, Wu, Long, Jiao, and Jin (2018), with permission.

REFERENCE, BIBLIOGRAPHY OR WORKS CITED

Ali, S., Singh, B., & Sharma, S. 2019. Impact of feed moisture on microstructure, crystallinity, pasting, physico-functional properties and in vitro digestibility of twin-screw extruded corn and potato starches. *Plant Foods for Human Nutrition* 74 (4): 474–480.

Ali, S., Singh, B., & Sharma, S. 2020. Effect of processing temperature on morphology, crystallinity, functional properties, and in vitro digestibility of extruded corn and potato starches. *Journal of Food Processing and Preservation* 44 (7): e14531.

Altomare, R. E., & Ghossi, P. 1986. An analysis of residence time distribution patterns in a twin screw cooking extruder. *Biotechnology Progress* 2 (3): 157–163.

Apruzzese, F., Pato, J., Balke, S. T., & Diosady, L. L. 2003. In-line measurement of residence time distribution in a co-rotating twin-screw extruder. *Food Research International* 36 (5): 461–467.

Berzin, F., Tara, A., Tighzert, L., & Vergnes, B. 2010. Importance of coupling between specific energy and viscosity in the modeling of twin screw extrusion of starchy products. *Polymer Engineering & Science* 50 (9): 1758–1766.

Bouvier, J. M., & Campanella, O. H. 2014. *Extrusion processing technology: food and non-food biomaterials.* New York: John Wiley & Sons.

Brown, S. B. 1991. Chemical processes applied to reactive extrusion of polymers. *Annual Review of Materials Science* 21: 409–435.

Brümmer, T., Meuser, F., van Lengerich, B., & Niemann, C. 2002. Effect of extrusion cooking on molecular parameters of corn starch. *Starch-Stärke* 54 (1): 1–8.

Calderón-Castro, A., Vega-García, M. O., de Jesús Zazueta-Morales, J., Fitch-Vargas, P. R., Carrillo-López, A., Gutiérrez-Dorado, R., Limón-Valenzuela, V., & Aguilar-Palazuelos, E. 2018. Effect of extrusion process on the functional properties of high amylose corn starch edible films and its application in mango (*Mangifera indica* L.) cv. Tommy Atkins. *Journal of Food Science and Technology* 55 (3): 905–914.

Chen, C. W., Don, T.-M., & Yen, H.-F. 2006. Enzymatic extruded starch as a carbon source for the production of poly(3-hydroxybutyrate-co-3-hydroxyvalerate) by *Haloferax mediterranei. Process Biochemistry* 41 (11): 2289–2296.

Chien, H.I., Tsai, Y.H., David Wang, H.M., Dong, C.D., Huang, C.Y., & Kuo, C.H. 2022. Extrusion puffing pretreated cereals for rapid production of high-maltose syrup. Food Chemistry X 15: 100445.

Chuang, C. C. G., & Yeh, A. I. 2004. Effect of screw profile on residence time distribution and starch gelatinization of rice flour during single screw extrusion cooking. *Journal of Food Engineering* 63 (1): 21–31.

Collazo-Bigliardi, S., Ortega-Toro, R., & Chiralt Boix, A. 2018. Reinforcement of thermoplastic starch films with cellulose fibres obtained from rice and coffee husks. *Journal of Renewable Materials* 6 (7): 599–610.

Cui, C., Ji, N., Wang, Y., Xiong, L., & Sun, Q. 2021. Bioactive and intelligent starch-based films: A review. *Trends in Food Science & Technology* 116: 854–869.

De Ruyck, H. 1997. Modelling of the residence time distribution in a twin screw extruder. *Journal of Food Engineering* 32 (4): 375–390.

Dircio-Morales, M. A., Fonseca-Florido, H. A., Velazquez, G., Ávila-Orta, C. A., Ramos-De Valle, L. F., Hernández-Gámez, F., Rivera-Salinas, J. E., & Soriano-Corral, F. 2022. Relationship among extrusion conditions, cell morphology, and properties of starch-based foams: A review. Starch-Stärke 75 (1–2): 2200103.

Duan, Q., Zhu, Z., Chen, Y., Liu, H., Yang, M., Chen, L., & Yu, L. 2022. Starch-based foams nucleated and reinforced by polysaccharide-based crystals. *ACS Sustainable Chemistry & Engineering* 10 (6): 2169–2179.

Duque, A., Manzanares, P., Ballesteros, I., Negro, M. J., Oliva, J. M., González, A., & Ballesteros, M. 2014. Sugar production from barley straw biomass pretreated by combined alkali and enzymatic extrusion. *Bioresource Technology* 158: 262–268.

Farahanchi, A., & Sobkowicz, M. J. 2017. Kinetic and process modeling of thermal and mechanical degradation in ultrahigh speed twin screw extrusion. *Polymer Degradation and Stability* 138: 40–46.

Fasheun, D. O., da Silva, A. S. A., Teixeira, R. S. S., & Ferreira-Leitão, V. S. 2023. Dark fermentative hydrogen production from cassava starch: A comprehensive evaluation of the effects of starch extrusion and enzymatic hydrolysis. *International Journal of Hydrogen Energy* 52: 322–334.

Ficarella, A., Milanese, M., & Laforgia, D. 2006. Numerical study of the extrusion process in cereals production: Part I. Fluid-dynamic analysis of the extrusion system. *Journal of Food Engineering* 73 (2): 103–111.

Froment, G. F., Bischoff, K. B., & Wilde, J. 1990. Chemical reactor analysis and design. New York: John Wiley.

Gandini, A., & Lacerda, T. M. 2015. From monomers to polymers from renewable resources: Recent advances. *Progress in Polymer Science* 48: 1–39.

Georges, A., Lacoste, C., & Damien, E. 2018. Effect of formulation and process on the extrudability of starch-based foam cushions. *Industrial Crops and Products* 115: 306–314.

Govindasamy, S., Campanella, O. H., & Oates, C. G. 1997. Enzymatic hydrolysis and saccharification optimisation of sago starch in a twin-screw extruder. *Journal of Food Engineering* 32 (4): 427–446.

Grimard, J., Dewasme, L., Thiry, J., Krier, F., Evrard, B., & Wouwer, A. V. 2016. Modeling, sensitivity analysis and parameter identification of a twin screw extruder. *IFAC-PapersOnLine* 49 (7): 1127–1132.

Gulzar, B., Hussain, S. Z., Naseer, B., & Naik, H. R. 2021. Enhancement of resistant starch content in modified rice flour using extrusion technology. *Cereal Chemistry* 98 (3): 634–641.

Huang, X., Liu, H., Ma, Y., Mai, S., & Li, C. 2022. Effects of extrusion on starch molecular degradation, order-disorder structural transition and digestibility-A review. *Foods* 11 (16): 2538.

Janssen, L. P. B. M. 2004. Reactive extrusion systems. New York: Marcel Dekker.

Kumar, A., Ganjyal, G. M., Jones, D. D., & Hanna, M. A. 2008. Modeling residence time distribution in a twin-screw extruder as a series of ideal steady-state flow reactors. *Journal of Food Engineering* 84 (3): 441–448.

Kuo, C.H., Shieh, C.J., Huang, S.M., David Wang, H.M., & Huang, C.Y. 2019. The effect of extrusion puffing on the physicochemical properties of brown rice used for saccharification and Chinese rice wine fermentation. *Food Hydrocolloids* 94: 363–370.

Lazou, A., & Krokida, M. 2010. Structural and textural characterization of corn-lentil extruded snacks. *Journal of Food Engineering* 100 (3): 392–408.

Lee, S. Y., & Mccarthy, K. L. 1996. Effect of screw configuration and speed on RTD and expansion of rice extrudate. *Journal of Food Process Engineering* 19 (2): 153–170

Levenspiel, O. 1972. Chemical reaction engineering. New York: John Wiley & Sons.

Li, B., Zhang, Y., Xu, F., Khan, M. R., Zhang, Y., Huang, C., Zhu, K., Tan, L., Chu, Z., & Liu, A. 2021. Supramolecular structure of artocarpus heterophyllus Lam seed starch prepared by improved extrusion cooking technology and its relationship with in vitro digestibility. Food Chemistry 336: 127716.

Li, J., Jiao, A., Rashed, M. M. A., Deng, L., Xu, X., & Jin, Z. 2018. Effect of thermostable α-amylase addition on producing the porous-structured noodles using extrusion treatment. *Journal of Food Science* 83 (2): 332–339.

Li, M., Hasjim, J., Xie, F., Halley, P. J., & Gilbert, R. G. 2014. Shear degradation of molecular, crystalline, and granular structures of starch during extrusion. *Starch-Stärke* 66 (7–8): 595–605.

Linko, P., Linko, Y. Y., & Olkku, J. 1983. Extrusion cooking and bioconversions. *Journal of Food Engineering* 2 (4): 243–257.

Liu, Q., Wang, Y., Yang, Y., Yu, X., Xu, L., Jiao, A., & Jin, Z. 2023. Structure, physicochemical properties and in vitro digestibility of extruded starch-lauric acid complexes with different amylose contents. *Food Hydrocolloids* 136: 108239.

Liu, W.C., Halley, P. J., & Gilbert, R. G. 2010. Mechanism of degradation of starch, a highly branched polymer, during extrusion. *Macromolecules* 43 (6): 2855–2864.

Liu, Y., Chen, J., Luo, S., Li, C., Ye, J., Liu, C., & Gilbert, R. G. 2017. Physicochemical and structural properties of pregelatinized starch prepared by improved extrusion cooking technology. *Carbohydrate Polymers* 175: 265–272.

Liu, Y., Chen, J., Wu, J., Luo, S., Chen, R., Liu, C., & Gilbert, R. G. 2019. Modification of retrogradation property of rice starch by improved extrusion cooking technology. *Carbohydrate Polymers* 213: 192–198.

Ma, S., Zhu, Q., Yao, S., Niu, R., Liu, Y., Qin, Y., Zheng, Y., Tian, J., Li, D., Wang, W., Liu, D., & Xu, E. 2022. Efficient retention and complexation of exogenous ferulic acid in starch: Could controllable bioextrusion be the answer? *Journal of Agricultural and Food Chemistry* 70 (47): 14919–14930.

Moad, G. 2011. Chemical modification of starch by reactive extrusion. *Progress in Polymer Science* 36 (2): 218–237.

Montilla-Buitrago, C. E., Gómez-López, R. A., Solanilla-Duque, J. F., Serna-Cock, L., & Villada-Castillo, H. S. 2021. Effect of plasticizers on properties, retrogradation, and processing of extrusion-obtained thermoplastic starch: A review. Starch-Stärke 73 (9–10): 2100060.

Offiah, V., Kontogiorgos, V., & Falade, K. O. 2018. Extrusion processing of raw food materials and by-products: A review. *Critical Reviews in Food Science and Nutrition* 59 (18): 2979–2998.

Pérez, S., & Bertoft, E. 2010. The molecular structures of starch components and their contribution to the architecture of starch granules: A comprehensive review. *Starch-Stärke* 62 (8): 389–420.

Qin, Y., Ma, S., Zhou, J., Li, D., Chen, J., Wang, W., Cheng, H., Wu, Z., Tian, J., Xu, E., & Liu, D. 2023. Conversion of endogenous phenolic acid in brown rice by bioextrusion of mesophilic α-amylase. Food Bioscience 53: 102699.

Shrestha, A. K., Ng, C. S., Lopez-Rubio, A., Blazek, J., Gilbert, E. P., & Gidley, M. J. 2010. Enzyme resistance and structural organization in extruded high amylose maize starch. Carbohydrate Polymers 80 (3): 699–710.

Sun, X., Sun, Z., Guo, Y., Zhao, J., Zhao, J., Ge, X., Shen, H., Zhang, Q., & Yan, W. 2021. Effect of twin-xuscrew extrusion combined with cold plasma on multi-scale structure, physicochemical properties, and digestibility of potato starches. Innovative Food Science & Emerging Technologies 74: 102855.

Tang, J., Zhou, J., Zhou, X., Li, D., Wu, Z., Tian, J., Xu, E., & Liu, D. 2022. Rearranged supramolecular structure of resistant starch with polymorphic microcrystals prepared in high-solid enzymatic system. Food Hydrocolloids 124(Part A): 107215.

Todd, D. B. 1993. Melting of plastics in kneading blocks. *International Polymer Processing* 8 (2): 113–118.

Torres-Aguilar, P. C., Hayes, A. M. R., Yepez, X., Martinez, M. M., & Hamaker, B. R. 2023. Formation of slowly digestible, amylose-lipid complexes in extruded wholegrain pearl millet flour. *International Journal of Food Science & Technology* 58 (3): 1336–1345.

Vergnes, B., & Berzin, F. 2004. Modelling of flow and chemistry in twin screw extruders. *Plastics, Rubber and Composites* 33 (9–10): 409–415.

Vergnes, B., & Berzin, F. 2006. Modeling of reactive systems in twin-screw extrusion: Challenges and applications. *Comptes Rendus Chimie* 9 (11–12): 1409–1418.

Vergnes, B., Souveton, G., Delacour, M. L., & Ainser, A. 2001. Experimental and theoretical study of polymer melting in a co-rotating twin screw extruder. *International Polymer Processing* 16 (4): 351–362.

Versino, F., Lopez, O. V., Garcia, M. A., & Zaritzky, N. E. 2016. Starch-based films and food coatings: An overview. *Starch-Stärke* 68 (11–12): 1026–1037.

Wang, N., Wu, L., Zhang, F., Kan, J., & Zheng, J. 2022. Modifying the rheological properties, in vitro digestion, and structure of rice starch by extrusion assisted addition with bamboo shoot dietary fiber. Food Chemistry 375: 131900.

Wang, R., Li, M., Chen, S., Hui, Y., Tang, A., & Wei, Y. 2019. Effects of flour dynamic viscosity on the quality properties of buckwheat noodles. *Carbohydrate Polymers* 207: 815–823.

Wang, R., Zhang, H., Chen, Z., & Zhong, Q. 2020. Structural basis for the low digestibility of starches recrystallized from side chains of amylopectin modified by amylosucrase to different chain lengths. Carbohydrate Polymers 241: 116352.

Wang, R., Zhang, T., He, J., Zhang, H., Zhou, X., Wang, T., Feng, W., & Chen, Z. 2019. Tailoring digestibility of starches by chain elongation using amylosucrase from *Neisseria polysaccharea* via a zipper reaction mode. *Journal of Agricultural and Food Chemistry* 68 (1): 225–234.

Wei, W., Wu, M., Ren, W., Yu, H., & Sun, D. 2024. Preparation of crosslinked starches with enhanced and tunable gel properties by the cooperative crosslinking-extrusion combined modification. Carbohydrate Polymers 324: 121473.

Wolf, D., & Resnick, W. 1963. Residence time distribution in real systems. *Industrial and Engineering Chemistry Fundamentals* 2 (4): 287–293.

Wolf, D., & White, D. H. 1976. Experimental study of the residence time distribution in plasticating screw extruders. *AICHE Journal* 22 (1): 122–131.

Wu, W., Jiao, A., Xu, E., Chen, Y., & Jin, Z. 2020. Effects of extrusion technology combined with enzymatic hydrolysis on the structural and physicochemical properties of porous corn starch. *Food and Bioprocess Technology* 13 (3): 442–451.

Xanthos, M. 1992. Reactive extrusion: principles and practice. New York: Hansen Publishers.

Xu, E., Campanella, O. H., Ye, X., Jin, Z., Liu, D., & BeMiller, J. N. 2020a. Advances in conversion of natural biopolymers: A reactive extrusion (REX)-enzyme-combined strategy for starch/protein-based food processing. *Trends in Food Science & Technology* 99: 167–180.

Xu, E., Li, D., Cheng, H., Zhao, H., Tian, J., Wu, Z., Chen, S., Ye, X., & Liu, D. 2020b. Effect of anion type on enzymatic hydrolysis of starch-(thermostable α-amylase)-calcium system in a low-moisture solid microenvironment of bioextrusion. Carbohydrate Polymers 240: 116331.

Xu, E., Pan, X., Wu, Z., Long, J., Li, J., Xu, X., Jin, Z., & Jiao, A. 2016. Response surface methodology for evaluation and optimization of process parameter and antioxidant capacity of rice flour modified by enzymatic extrusion. *Food Chemistry* 212: 146–154.

Xu, E., Wu, Z., Chen, J., Tian, J., Cheng, H., Li, D., Jiao, A., Ye, X., Liu, D., & Jin, Z. 2020c. Calcium-lactate-induced enzymatic hydrolysis of extruded broken rice starch to improve Chinese rice wine fermentation and antioxidant capacity. Lwt 118: 108803.

Xu, E., Wu, Z., Ding, T., Ye, X., Jin, Z., & Liu, D. 2019. Magnetic (Zn-St)10FeOn (n=1, 2, 3, 4) framework of macro-mesoporous biomaterial prepared via green enzymatic reactive extrusion for dye pollutants removal. *ACS Applied Materials & Interfaces* 11 (46): 43553–43562.

Xu, E., Wu, Z., Jiao, A., & Jin, Z. 2018a. Effect of exogenous metal ions and mechanical stress on rice processed in thermal-solid enzymatic reaction system related to further alcoholic fermentation efficiency. *Food Chemistry* 240: 965–973.

Xu, E., Wu, Z., Jin, Z., & Campanella, O. H. 2017a. Bioextrusion of broken rice in the presence of divalent metal salts: Effects on starch microstructure and phenolics compounds. *ACS Sustainable Chemistry & Engineering* 6 (1): 1162–1171.

Xu, E., Wu, Z., Li, J., Pan, X., Sun, Y., Long, J., Xu, X., Jin, Z., & Jiao, A. 2017b. Residence time distribution for evaluating flow patterns and mixing actions of rice extruded with thermostable α-amylase. *Food and Bioprocess Technology* 10 (6): 1015–1030.

Xu, E., Wu, Z., Long, J., Jiao, A., & Jin, Z. 2018b. Porous starch-based material prepared by bioextrusion in the presence of zinc and amylase-magnesium complex. *ACS Sustainable Chemistry & Engineering* 6 (8): 9572–9578.

Xu, E., Wu, Z., Wang, F., Li, H., Xu, X., Jin, Z., & Jiao, A. 2015. Impact of high-shear extrusion combined with enzymatic hydrolysis on rice properties and Chinese rice wine fermentation. *Food and Bioprocess Technology 8* (3): 589–604.

Yan, X., Wu, Z.-Z., Li, M.-Y., Yin, F., Ren, K.-X., & Tao, H. 2019. The combined effects of extrusion and heat-moisture treatment on the physicochemical properties and digestibility of corn starch. *International Journal of Biological Macromolecules* 134: 1108–1112.

Yang, Q., Yang, Y., Luo, Z., Xiao, Z., Ren, H., Li, D., & Yu, J. 2016. Effects of lecithin addition on the properties of extruded maize starch. *Journal of Food Processing and Preservation* 40 (1): 20–28.

Ye, J., Hu, X., Luo, S., Liu, W., Chen, J., Zeng, Z., & Liu, C. 2018. Properties of starch after extrusion: A review. Starch-Stärke 70 (11–12): 1700110.

Yeh, A. I., & Jaw, Y. M. 1999. Predicting residence time distributions in a single screw extruder from operating conditions. *Journal of Food Engineering* 39 (1): 81–89.

Yu, L., Meng, Y., Ramaswamy, H. S., & Boye, J. 2014. Residence time distribution of soy protein isolate and corn flour feed mix in a twin-screw extruder. *Journal of Food Processing and Preservation* 38 (1): 573–584.

Zeller, B. L., Saleeb, F. Z., & Ludescher, R. D. 1998. Trends in development of porous carbohydrate food ingredients for use in flavor encapsulation. *Trends in Food Science & Technology* 9: 389–394.

Zhai, X., Wang, W., Zhang, H., Dai, Y., Dong, H., & Hou, H. 2020. Effects of high starch content on the physicochemical properties of starch/PBAT nanocomposite films prepared by extrusion blowing. Carbohydrate Polymers 239: 116231.

Zhang, B., Dhital, S., Flanagan, B. M., Luckman, P., Halley, P. J., & Gidley, M. J. 2015. Extrusion induced low-order starch matrices: Enzymic hydrolysis and structure. *Carbohydrate Polymers* 134: 485–496.

Zhang, Z., Zhu, M., Xing, B., Liang, Y., Zou, L., Li, M., Fan, X., Ren, G., Zhang, L., & Qin, P. 2023. Effects of extrusion on structural properties, physicochemical properties and in vitro starch digestibility of Tartary buckwheat flour. *Food Hydrocolloids* 135: 108197.

Zhao, S., Jiao, A., Yang, Y., Liu, Q., Wu, W., & Jin, Z. 2022. Modification of physicochemical properties and degradation of barley flour upon enzymatic extrusion. *Food Bioscience* 45: 101243.

Zheng, Y., Tian, J., Ogawa, Y., Yin, X., Xu, E., Chen, S., Liu, D., Kong, X., & Ye, X. 2021. Co-extrusion of proanthocyanins from Chinese bayberry leaves modifies the physicochemical properties as well as the in vitro digestion of restructured rice. *Food Structure* 27: 100182.

CHAPTER 5

Electric Field Treatment of Starch and Starchy Materials

Dandan Li, Na Yang, and Yanfeng Ding

5.1 INTRODUCTION

In order to widen the applications of native starch, various modification procedures have been applied to improve its physicochemical and functional properties. Along with the increasing consumer demands for healthy products, more attention is being paid to physical technologies that are environmentally friendly (Grgić et al., 2019). Electric field-based techniques are green and efficient in altering the microstructure and macromolecular interactions of biomacromolecules (Giteru, Oey, & Ali, 2018). Since Varella et al. (1984) first proposed continuous enzymatic cooking and liquefaction of starch by using direct resistive heating (ohmic heating, OH) to replace traditional water bath or steam heating. Then, various researches related to the electro-accelerated modification of starches were reported (Figure 5.1).

Compared with traditional heating treatment, OH and induced electric field (IEF) could achieve rapid heat and mass transfer while improving reaction selectivity by accelerating the directional migration of charged species (Li et al., 2021; Wang & Sastry, 1997); pulsed electric field (PEF) could increase the accessibility of starch chains to enzyme and chemical reagents by destroying the densely packed semicrystalline within starch

DOI: 10.1201/9781003493594-5

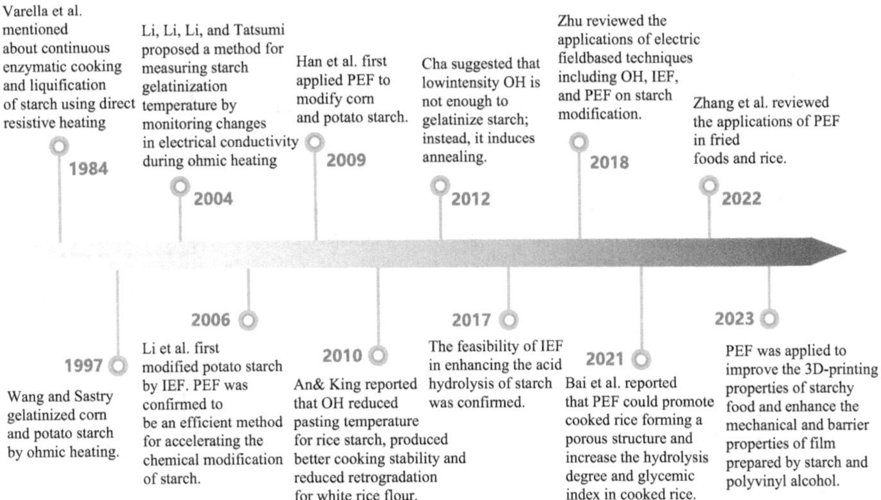

FIGURE 5.1 Timeline of important events/researches in the field of electric field treatment of starch.

granules (Hong et al., 2016; Zhu, 2018). Moreover, some researchers believe that the catalytic process of enzymes is essentially a powerful electrostatic catalytic process. If a high-strength electrostatic field similar to that of an enzyme-catalytic process can be generated in practice, it may be possible to simulate the catalytic activity of the enzyme (Fried, Bagchi, & Boxer, 2014). From this aspect, exploring the applications of electric field in biopolymer modification has obvious strategic significance. Therefore, this chapter introduced the working principle of electric field-based techniques, clarified the mechanism of electric action on starch granules, discussed the structural change of electric field-treated starch, and summarized the potential applications of electric field-treated starch in industries.

5.2 ELECTRIC FIELD-BASED TECHNIQUES AND PRINCIPLES

The electric field-based techniques used for starch modifications include OH, PEF, and IEF (Figure 5.2). OH and PEF are produced by imposing electric voltage on metal electrodes, while IEF is induced by electromagnetic conversion in the configuration of a transformer.

5.2.1 Ohmic Heating

OH is an electroheating technique on the basis of passing electric current through a product having electrical resistance (Goullieux & Pain, 2014; Icier, 2012). Metal electrodes are commonly used as the electrical

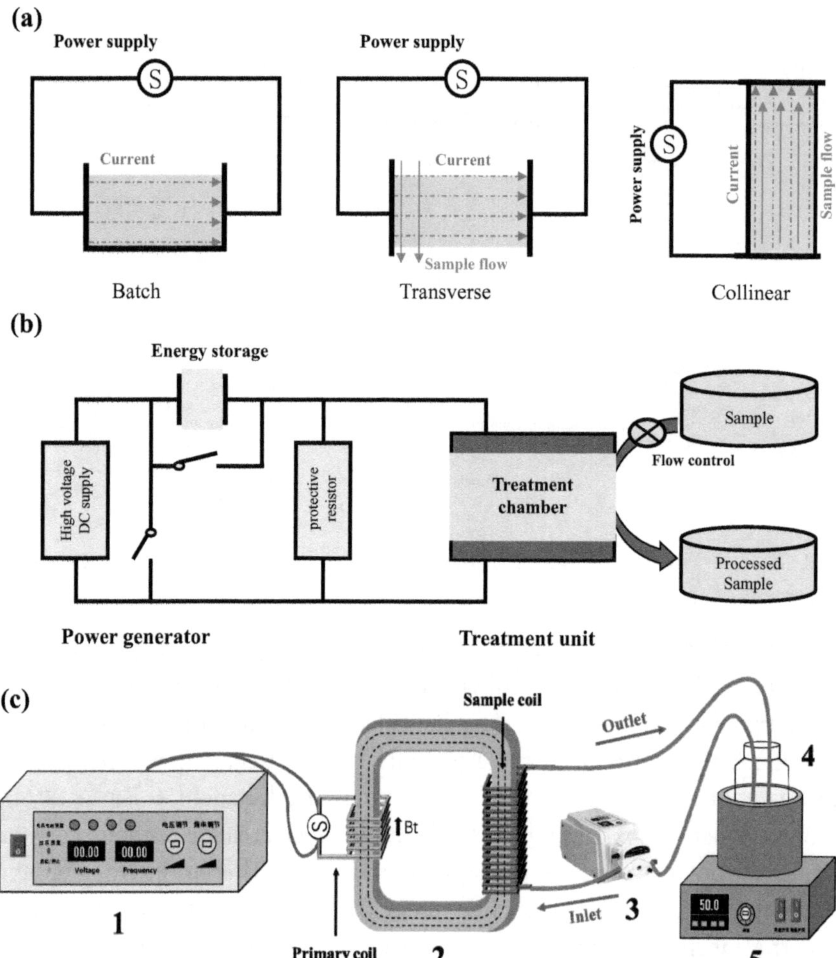

FIGURE 5.2 Schematic figures of (a) ohmic heating, (b) pulsed electric field, and (c) induced electric field (Li, Liu, Tao, Shi, Wang, & Han, 2023).

conductor, while the treated sample is the conductive medium. The metal electrodes, which are in intimate contact with the sample, are separated by an electrically insulated tube or plate space. Goullieux and Pain (2014) summarized three common generic configurations (batch, transverse, and collinear) of OH systems (Figure 5.2a). For batch configuration, the electrodes are coaxial, or plane and parallel; for transverse configuration, the electrodes are generally plane or coaxial and slightly spaced, while the product flows parallel to the electrodes and perpendicularly to the electric field; for collinear configuration, the electrodes are generally widely spaced, while the product fluid flows from one electrode to another. During

OH treatment, metal electrodes are fed by an electric power supply, which is direct, alternating, or pulsed. The imposed electric current then drives the directional migration of charged species in the treated sample with the primary purpose of heating it.

5.2.2 Pulsed Electric Field

PEF is a non-thermal technique that involves the application of short pulses (μs-ms) of the high-voltage electric field to samples placed between two electrodes (Giteru et al., 2018). A typical PEF system contains a power generator and a treatment chamber (Figure 5.2b). The power generator is a combination of any passive discrete elements (capacitive, inductive, and resistive), transformers, and power switches. The treatment chamber, which is batch and continuous, comprises two electrodes (parallel, coaxial, and colinear are the typically used configurations) held in a particular position by insulators that also form an enclosure containing conducting materials (Giteru et al., 2018). During the PEF treatment, the pulse power generator delivers the pulsating high-voltage across the treatment chamber containing samples, thus contributing to the changes in the structural and functional properties of biomacromolecules. Square and exponentially decaying pulses in either monopolar or bipolar forms are two commonly applied pulse waveforms (Arshad et al., 2020).

5.2.3 Induced Electric Field

IEF is a non-electrode contacting electro-technique, which is developed *via* electromagnetic induction (Li et al., 2023) (Figure 5.2c). It consists of (1) a power source, (2) an IEF reactor, (3) a peristaltic pump, (4) a sample tank, and (5) a circulating water bath, among which the IEF reactor is developed in the configuration of a transformer with high-frequency power, which is composed of (6) a primary coil, (7) one or several secondary coils, and (8) a magnetic core. Differing from traditional transformers, the secondary coils of IEF were formed by a liquid conductor constrained by thin and non-conductive tubing, which is also called sample coils. According to Faraday's law, an alternating magnetic flux will be induced in the magnetic core when an alternating excitation voltage is imposed on the primary coil (Pryor, 2013). Then, the magnetic flux induces an alternating electric voltage in sample coils, which drives the migration of charged species in liquid sample and thus enhances mass and heat transfer. The effect of IEF on starch granules was similar to that of alternating OH, due to the same range of electric field intensity and frequency applied.

5.3 EFFECT OF ELECTRIC ACTION ON STARCH

5.3.1 Thermal Effect

When electric current passes through samples having finite conductivity, electrical energy converts to heat due to the transfer of energy from conducted electrons to the conductor's atoms by collisions (Redondo et al., 2018) (Figure 5.3a). According to Joule's law, the amount of heat is proportional to the square of electric field intensity and the sample conductivity (Knirsch et al., 2010). The increased temperature will induce the partial or full gelatinization of starch granules and also lead to changes in the efficiency of starch chemical/enzymatic modification by significantly affecting the activation energy of the reaction system or altering the enzyme activity. Traditional heating techniques require a temperature gradient to transfer heat to the treated sample, which causes the surfaces presenting a higher temperature than the interior (Sakr & Liu, 2014). For starch or starchy materials, starch granules absorb water, swell, disassemble, and finally form a high-viscous paste in the presence of excess water (Li et al., 2022). This high viscosity hinders the heat transfer during traditional heating, leading to the fouling of surfaces which become burnt onto the hot surfaces adversely affecting the product. OH and IEF, however, provide rapid and uniform heating by internal energy transformation (from electric to thermal) within the sample. The absence of a hot surface reduces fouling problems and decreases thermal damage to products. Although PEF is typically considered a non-thermal technique due to the short pulses and low power applied, some researchers believed that the destruction of the semicrystalline and granular structure of starch caused by PEF was attributed to the surface gelatinization (Chen et al., 2021; Hong et al., 2016). Han et al. (2012) reported that the temperature elevation caused by 30, 40, and 50 kV/cm PEF treatment was 4.7°C, 6.3°C, and 7.3°C, respectively.

5.3.2 Non-thermal Effect

Besides thermal effect, the external electric field can also directly change the microscopic motion of particles such as electrons, atoms, and molecules, thereby affecting the macroscopic mass transfer and the progress of reactions (Figure 5.3a). Pinto et al. (2013) suggested that OH allows for higher reaction yields and shorter reaction times for four representative organic transformations by inducing faster and more uniform heating and an increase of dynamics/mobility of charged species, when compared with those obtained under oil bath and microwave heating. Electric field can also cause molecular polarization by inducing internal charge flow (Figure 5.3b). The positive and negative charges inside polar molecules

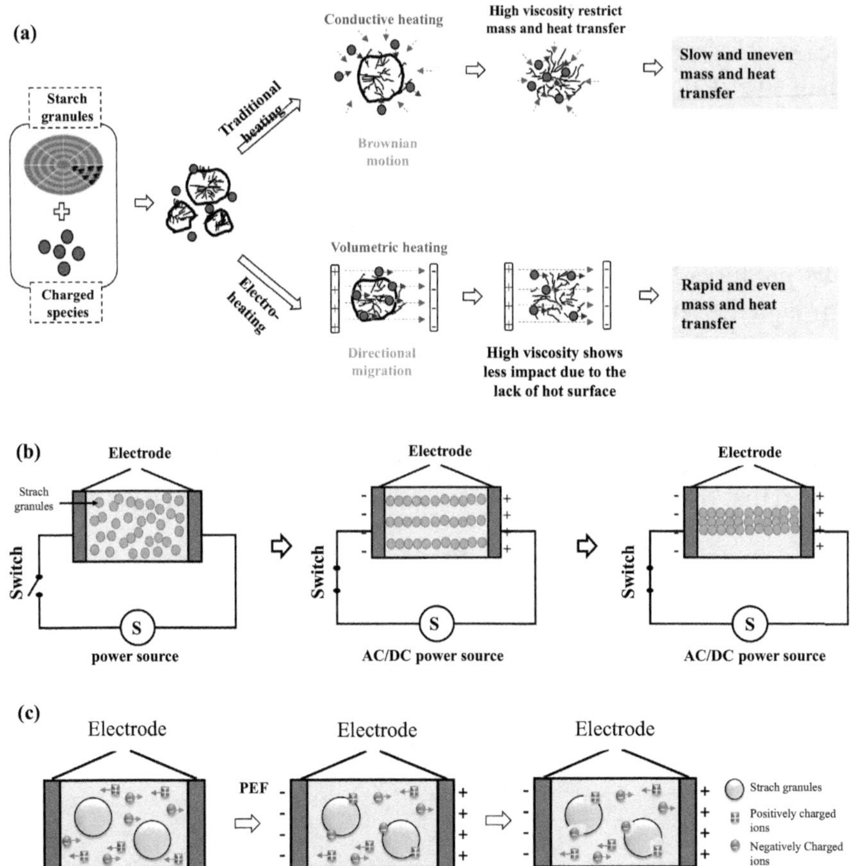

FIGURE 5.3 (a) Mass and heat transfer during traditional and electro-heating; (b) electro-driven polarization of starch granules; and (c) pulsed electric field acting on starch granules by producing space polarization charges.

shift to the opposite direction, resulting in the presence of electric dipole distance and displacement polarization. Meanwhile, due to the uneven distribution of charges in polar molecules, their inherent dipole distances tend to align along the direction of the electric field, leading to directional polarization. The self-assembly of starch and its derivatives such as phosphate starch and kaolinite/carboxymethyl starch hybrid dispersed in insulating liquids driven by electric field has been widely reported (Sung et al., 2005; Wang & Zhao, 2002). In addition, charged ions such as K^+ and Cl^- will move and gather on the surface of starch granules under PEF, forming space polarization charges (Figure 5.3c). With the increase of PEF intensity, an instantaneous high-voltage discharge is generated on the outer layer of starch granules, causing roughness and disruption. Han et al. (2012)

suggested that some roughness or damages emerged on the surface of tapioca starch granules when subjected to PEF treatment at intensities higher than 30 kV/cm.

5.4 STRUCTURAL CHANGE OF ELECTRIC FIELD-TREATED STARCH (EFTS)

5.4.1 Ohmic Heating on Starch Structure

OH produces heat rapidly and uniformly by passing the electric current through the sample. The heat generated by low-intensity OH is not enough to gelatinize starch; instead, it induces annealing (Cha, 2012) (Figure 5.4a), which is defined as a process that treats starch at a temperature above the glass transition but below the gelatinization temperature in the presence of excess water (Zavareze & Dias, 2011). Jayakody and Hoover (2008) suggested that the annealing of semicrystalline polymers could be interpreted

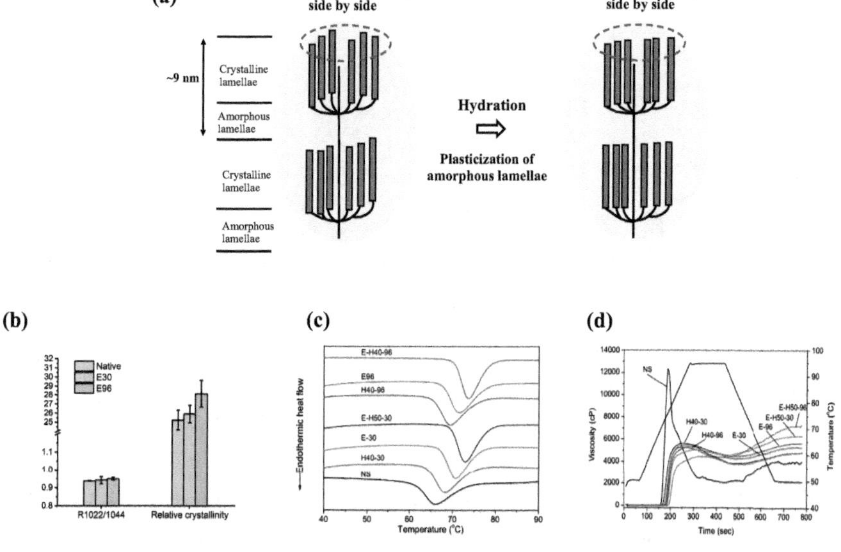

FIGURE 5.4 Changes in structural and physicochemical characteristics of starch under low-intensity ohmic heating (OH) and induced electric field (IEF) treatment: (a) Possible mechanism; Effect of IEF treatment on the (b) ratio of FTIR peak intensity at 1044 and 1022 cm^{-1} and relative crystallinity, (c) differential scanning calorimetry curve, and (d) rapid viscosity curves of potato starch (Li et al., 2016). Note: E30 and E96 represent IEF treatment for 30 and 96 hours, respectively; H40-30 and H40-96 represent heating at 40°C for 30 and 96 hours, respectively; E-H50-30 and E-H50-96 represent a combination of IEF treatment and heating at 50°C for 30 and 96 hours, respectively.

as: (1) "sliding diffusion", where complete molecular sequences move within a crystalline lattice, and/or (2) "complete or partial fusion" of the crystals, followed by a subsequent re-crystallization of the melted polymers at the annealing temperature. After annealing, the crystalline perfection is improved and the interactions between starch chains are facilitated (Zavareze et al., 2011).

When the heat energy produced by OH obtains a critical value, starch granules start to absorb water, swell, and then disrupt to form a high-viscous paste (Figure 5.3a). Compared with traditional cooking, ohmic cooking saved more than 70% of energy; in addition, no fouling (the rice layer) was observed on the cooking container after ohmic cooking because no heating surface is needed in OH (Kanjanapongkul, 2017). The effect of OH on starch gelatinization depends on the starch type, electric field intensity, electrical conductivity, etc. The degree of gelatinization was 70% for jicama starch, while it was only 39% for cassava starch, after OH treatment at 123 V for 10 minutes (Martínez-Bustos et al., 2005). The heating rate of starch suspension increased from 4 to 61°C/min, with an increasing OH intensity from 20 to 70 V/cm (An & King, 2010). Kaur and Singh (2015) reported that the heating rate of OH is proportional to the square of electric field strength and electrical conductivity. Starches with diverse degrees of gelatinization showed different functional properties, which could meet the different applications in industries.

5.4.2 Pulsed Electric Field on Starch Structure

PEF modification of starch involves the application of electrical pulses with high voltage (1–80 kV/cm) to starch suspensions or pastes for short durations (μs-ms) (Castro et al., 2023). This process causes physical and chemical changes in the starch molecules, including damage to starch granules, increases in granular size, changes in relative crystallinity, and alteration of molecular weight (Figure 5.5) (Almeida et al., 2022; Gulzar et al., 2023). The damage to starch granules caused by low-intensity PEF is dependent on the botanical origin. PEF treatment ranging from 2.86 to 8.57 kV/cm did not induce any damage on the granular morphology of starches from wheat, potato, and pea (Li et al., 2019), whereas sunken areas and fractures presented on the surface of rice starch granules when subjected to same PEF intensities (Wu et al., 2019). Higher-intensity PEF appears to cause more evident damages to the granular morphology of starch (Han et al., 2012; Han et al., 2009a, b; Zeng et al., 2016). PEF treatment at 30 kV/cm induced roughness on starch granules; at 40 kV, some pits were formed; and

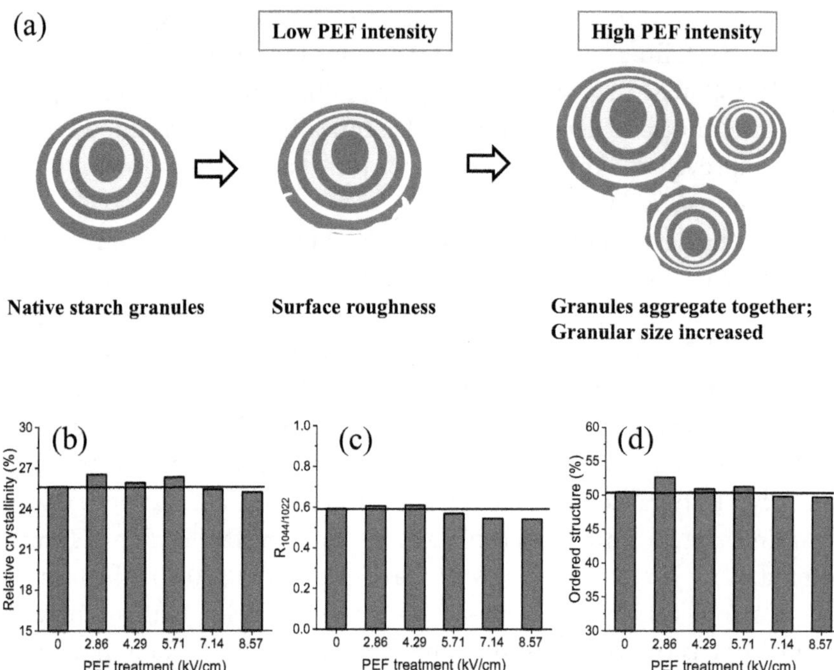

FIGURE 5.5 Effect of pulsed electric field (PEF) treatment on the (a) granular morphology and size, (b) relative crystallinity, (c) ratio of FTIR peak intensity at 1044 and 1022 cm^{-1}, and (d) ratio of the ordered structure determined by Roman spectra of starch. Note: Figure 5.5b and c was drawn according to the reports of Han, Zeng, Yu, et al. (2009).

when the intensity is up to 50 kV/cm, some starch granules were twisted and flocked and severe aggregation of starch granules occurred. As a result of surface disruption and aggregation, the granular size of PEF-treated starches significantly increased. Han et al. (2009a) revealed that the volume-mean diameter of potato starch gradually increased from 37.93 μm (0 kV/cm) to 56.11 μm (30 kV/cm), 85.16 μm (40 kV/cm), and 113.84 μm (50 kV/cm) after the PEF treatment.

Starch granules are formed in concentric alternating semicrystalline and amorphous layers (Svihus, Uhlen, & Harstad, 2005). The short-range molecular order and the double-helix packing within starch granules could be reflected by the intensity ratio of bands of Fourier transform infrared spectra at ~1044 and ~1022 cm^{-1} ($R_{1044/1022}$) and the C1 and C4 positions in nuclear magnetic resonance spectra, while the changes in crystallinity of long-range molecular order could be determined by polarized light microscope and X-ray diffractometry (Castro et al., 2023; Svihus et al., 2005). PEF can change the radial arrangement of amylopectin in semicrystalline layers, which depends on the starch type and PEF intensity (Castro et al., 2023). During low-intensity PEF treatment, the input energy is not enough to disrupt the arrangement of starch chains; instead, it allows for the reorganization of amylose chains in loosely packed amorphous areas, thus leading to an increase of relative crystallinity, $R_{1044/1022}$, and ordered structure. Li et al. (2019) reported that the PEF treatment at 2.86–4.29 kV/cm did not induce the vanishment of birefringence and diffraction peaks in wheat, pea, and potato starches, but the relative crystallinity, $R_{1044/1022}$, and order structure calculated by ^{13}C NMR spectra slightly increased. Wu et al. (2019) also suggested that the PEF treatment could slightly increase the relative crystallinity, $R_{1044/1022}$, and order structure of waxy rice starch. However, during PEF treatment with higher intensity, more energy is provided to disrupt the non-covalent bonds such as hydrogen bonding and van der Waals force among starch chains, which leads to the transformation of starch granules from crystal into non-crystal and amorphous and in consequence weakens the diffraction peaks, decreases the relative crystallinity, and even induces the breakage of starch chains. The Maltese crosses of some wheat and potato starch granules faded after PEF treatment at 7.14 and 8.57 kV/cm; in addition, the order structure of the wheat and potato starches was decreased by 0.7% and 1.6%, respectively, while the $R_{1044/1022}$ of waxy rice starch was significantly reduced by 0.012 after PEF treatment at 8.75 kV/cm (Li et al., 2019). Han et al. (2009a, b; 2012) and Zeng et al. (2016) also revealed that the diffraction peak intensity of waxy rice, tapioca, corn, and potato starches was weakened, the relative crystallinity was decreased, and the molecular weight was reduced, with the increasing of PEF intensity from 30 to 50 kV/cm.

5.4.3 Induced Electric Field on Starch Structure

The field intensity and frequency of IEF are in the same range as alternating-current OH, and hence their actions on starch granules are also similar. Low-intensity IEF causes the annealing, while high-intensity induces the partial or full gelatinization of starch. Electric field can drive the self-assembly of starch chains (Sung et al., 2005; Wang et al., 2002), and thus might help better pack the double helices within starch semicrystalline. Li et al. (2016) claimed the enhanced crystalline perfection to IEF enhancing the associations among starch chains and their reorientation. However, an insulating dispersion (typically, silicone oil) and a high-intensity electric field (~kV) are commonly necessary for the occurrence of assembly and alignment of polymer chains. The field intensity (15–75 V) applied in the research of Li et al. (2016) was relatively low. Notably, the IEF processing was conducted at 40°C and 50°C, at which annealing is easy to occur (Figure 5.4). Hence, the improved ordered structure of starch granules during low-intensity IEF treatment might be more likely attributed to starch annealing rather than IEF-induced self-assembly. Meanwhile, IEF can drive the migration of charged species to form ion current, thus producing Joule's heat. As the excitation voltage and electrical conductivity increase, the heat energy produced by IEF increases. Li et al. (2018; 2020) monitored the temperature changes during IEF-assisted acid hydrolysis of polysaccharides and found an enhanced thermal effect for IEF treatment with higher excitation voltage and higher salt concentration. Li et al. (2023) found that partial gelatinization might occur on the surface of starch granules after the IEF treatment at an excitation voltage of 400 V, which resulted in the aggregation of starch granules when dispersed in water.

5.5 EFTS APPLICATIONS

This section summarizes the applications of electric field-based techniques including PEF, OH, and IEF in starch and starchy materials. OH and IEF can impose the rapid migration of charged species, thus creating a range of starch functionalities by enhancing mass and heat transfer. Low-intensity OH and IEF cause the annealing of starch, which typically increases its gelatinization temperature, narrows its gelatinization temperature range, and thus exhibits various applications in industries such as packaging materials and additives. High-intensity OH and IEF tend to gelatinize starch, showing excellent potential applications as a thickening and gelling agent in frozen/instant foods or heat-sensitive products, as well as a binder and a disintegrant in the production of rapid orally disintegrating tablets in pharmaceutical industries. PEF is a non-thermal technique that involves applications of high-voltage pulses to starch and starchy materials for short durations (μs-ms). It could cause damage to starch granules, alter its semicrystalline perfect, and decrease its molecular weight, thus being employed to increase the digestibility and enhance rice qualities such as flavor and texture quality, reduce oil uptake and alter the texture of fries, improve the 3D-printing properties, and improve the mechanical and barrier properties of starch-based biodegradable films. Overall, electric field-based techniques could be efficient and green in improving the structural and physicochemical properties of starch and starchy materials.

5.5.1 Packaging Materials and Food/Non-Food Additives by Ohmic Heating and Induced Electric Field

OH and IEF could induce the annealing or gelatinization of starch, which mainly depends on electric field intensity, frequency, electrical conductivity, starch type, etc. (Table 5.1). Typically, low-intensity electric field treatment of starches at sub-gelatinization temperatures causes annealing (Cha, 2012; Li et al., 2016). Modification of starch by annealing affects its structural and physicochemical characteristics, including crystallinity, water solubility, swelling power, pasting and gelatinization properties, and thermal and freeze-thaw stability (Tester & Debon, 2000; Zavareze et al., 2011), thus exhibiting various industrial applications as packaging materials and additives (Fonseca et al., 2021). Li et al. (2016) exposed IEF (excitation voltage of 0–75 V) to potato starch suspensions for 30–96 hours, and found a better-organized structure was obtained for IEF-treated potato starch, proved by the increased ratio of 1044/1015 cm^{-1} and relative crystallinity (Figure 5.4b). The enhanced crystalline order further increased the gelatinization temperatures, narrowed the range of the gelatinization temperature, lowered the peak viscosity, and restricted the short-term retrogradation, which expanded the applications of native starch (Figure 5.4c and d). Cha (2012) suggested that OH appeared to be more effective in starch annealing when compared with conventional heating.

When the temperature during OH and IEF treatment obtains the onset gelatinization temperature, starch starts to absorb water, swell, and then gelatinize. Pre-gelatinization is a common physical modification of starch, while pre-gelatinized starch, which is also called “instant” starches, can be dissolved in cold water and shows good thickening/gelling capabilities (BeMiller, 2016). At present, fully pre-gelatinized starches have been widely applied as a thickening and gelling agent in frozen/instant foods or heat-sensitive products, such as cold desserts, salad dressings, bakery mixes, and baby foods (Liu & Hong, 2018). Partially pre-gelatinized starch, which has a mixture of properties of both native and fully gelatinized starches, has been successfully used as both a binder and a disintegrant in the production of rapid, orally disintegrating tablets in pharmaceutical industries over the last decades (Alalaiwe et al., 2019; Mimura et al., 2011). Conventional methods for starch gelatinization are based on heat transfer, potentially lengthening the heating duration with the possible overprocessing of individual product fractions caused by the high viscosity of the starch paste. OH and IEF are alternative thermal methods without long heating duration, overprocessing, and unwanted temperature peaks due to the direct dissipation of electrical

TABLE 5.1 Effect of Ohmic Heating (OH) and Induced Electric Field (IEF) Treatment on the Structural and Physicochemical Characteristics of Starch and Starchy Materials and the Potential Applications of OH-Treated and IEF-Treated Starch Samples

Electro-Techniques	Electric Field Treatment Conditions	Starch and Starchy Materials	Results	Possible Applications	References
OH	Low-intensity field; temperature remains below the onset gelatinization of starches	Potato starch, wheat starch, corn starch, sweet potato starch	Annealing occurred and OH seemed to cause more extensive annealing than traditional water-bath heating. After OH treatment, the gelatinization temperatures of starches increased but the temperature range narrowed. The variation degree showed the following sequence: potato starch > wheat starch > corn starch > sweet potato starch.	Biodegradable films; noodles; bakery foods; canned and frozen goods; and functional food products.	Cha (2012, 2014)
IEF		Potato starch	Annealing occurred. After IEF treatment at an excitation voltage of 75 V, the short-range and long-range order structure was improved, the gelatinization temperature and gelatinization enthalpy increased, the gelatinization temperature range narrowed, the swelling power, pasting viscosities, and setback value decreased.		Li et al. (2016)

(*Continued*)

TABLE 5.1 (*Continued*) Effect of Ohmic Heating (OH) and Induced Electric Field (IEF) Treatment on the Structural and Physicochemical Characteristics of Starch and Starchy Materials and the Potential Applications of OH-Treated and IEF-Treated Starch Samples

Electro-Techniques	Electric Field Treatment Conditions	Starch and Starchy Materials	Results	Possible Applications	References
OH	High-intensity field; temperature exceeds the onset gelatinization of starches	Corn starch	As the temperature during OH increased to 62°C, starch started to gelatinize, and the viscosity of starch suspension dramatically increased, which restricted mass and heat transfer, and thus slowed down the increase rate of conductivity	Using as a thickening and gelling agent in frozen/instant foods or heat-sensitive products, such as cold desserts, salad dressings, bakery mixes, and baby foods.	Karapantsios, Sakonidou, & Raphaelides (2010)
		Corn starch, mung bean starch, potato starch	OH treatment at 10 V/cm and 50 Hz could cause the gelatinization of starches		Li et al. (2004)
		Cassava and jicama starches	Starch with different degrees of gelatinization could be obtained by monitoring OH treatment conditions. Greater degrees of gelatinization were obtained by OH with higher voltage and short processing time (123 V, 10 minutes). At this condition, the gelatinization for jicama and cassava starches was 70.0% and 39.1%, respectively.	Using as both a binder and a disintegrant in the production of rapid orally disintegrating tablets in pharmaceutical industries over the last decades.	Martínez-Bustos et al. (2005)

(*Continued*)

TABLE 5.1 (*Continued*) Effect of Ohmic Heating (OH) and Induced Electric Field (IEF) Treatment on the Structural and Physicochemical Characteristics of Starch and Starchy Materials and the Potential Applications of OH-Treated and IEF-Treated Starch Samples

Electro-Techniques	Electric Field Treatment Conditions	Starch and Starchy Materials	Results	Possible Applications	References
		Rice	OH was proposed as an alternative method to cook rice. The applied electric field facilitated water diffusion, thus accelerating the cooking efficiency. Compared with a commercial electric rice cooker, the cooking energy was saved more than 70%		Kanjanapongkul (2017)
		Rice starch and rice flour.	As the OH intensity increased from 20 to 70 V/cm, the time for starch suspension to obtain 100°C decreased from 20 minutes to 78 seconds. Compared with traditional water-bath heating, OH-treated rice flour showed a better cooking stability, a reduced setback value, and decreased gelatinization temperatures.		An et al. (2010)
IEF		Maize starch	Partial gelatinization occurs after IEF treatment at an excitation voltage of 400 V. The surface of starch granules becomes rough, and some granules aggregate together when dispersed in water.		Li et al. (2023)

energy into Joule heat (Jaeger et al., 2016; Yang et al., 2017). An et al. (2010) reported that both OH and conventional heating led to starch pre-gelatinization, but the degree of gelatinization of OH-treated starch was higher.

5.5.2 Quality Modification of Starch Products by Pulsed Electric Field

PEF exhibits various applications in starch and starchy materials, which is summarized in Figure 5.6. PEF has gained increasing attention in improving the 3D-printing properties in recent years. PEF at high electric field intensities typically facilitates the swelling of starch granules, causes instantaneous starch gelatinization, and promotes the formation of gels with high stability, which allows for a greater scope for PEF-treated starches to be used in 3D-printing applications (Gulzar et al., 2023). Maniglia et al. (2021) treated wheat and cassava starches at 15–25 kV/cm electric field strengths and energy inputs and found that 25 kV/cm PEF-modified wheat starch outperformed the 3D-printing properties, including higher hardness, lower adhesiveness, and smoother surface compared with native starch, whereas PEF-treated cassava showed no significant changes. PEF was also reported to improve the mechanical and barrier qualities of starch-based biodegradable films. The film prepared by PEF-modified elephant foot yam starch and polyvinyl alcohol showed increased tensile strength, elongation percentage, and moisture barrier properties (Singh et al., 2023). Furthermore, PEF has also been used as a pretreatment method to enhance the physicochemical properties of

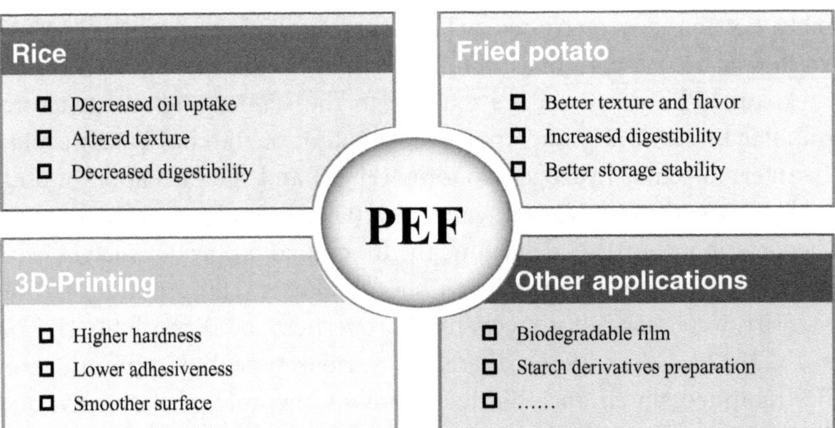

FIGURE 5.6 Applications of pulsed electric field (PEF) in starch and starchy materials.

chemically modified starches such as dialdehyde starch, porous starch, oxidized starch, and so on (Li et al., 2023a, b; Luo et al., 2023). These modified starches showed various industrial applications in food, pharmaceuticals, textiles, and other fields.

PEF can also applied to improve the quality of starchy food. Bai et al. (2021) and Qiu, Abbaspourrad, and Padilla-Zakour (2021; 2022) reported that PEF could create micro-pores on the surface of whole, water-soaked rice grains, thus allowing rice samples to retain more moisture, and better diffusion of sugar, oil and gum; moreover, PEF could also decrease the crystalline degree, reduce the gelatinization enthalpy, and decrease the peaking viscosity of rice flour. At present, PEF has been successfully employed to improve the texture, flavor, digestibility, and storage stability of cooked rice and rice-based products (Zhang et al., 2022). Another use of PEF in starchy food is to reduce the oil content, significantly change the textural properties, and decrease the digestibility of fried potato samples in the form of cubes, slices, crisps, strips, or sticks (Leong et al., 2022; Zhang et al., 2022). Fried foods are popular worldwide due to their peculiar organoleptic properties. However, the high fat content and calories increase the risk of adverse health consequences. PEF was reported to cause cracks and collapse in the cell walls of potato slices, which accelerated the evaporation of water and simultaneously reduced oil absorption during the frying process (Ignat et al., 2015; Janositz, Noack, & Knorr, 2011; Zhang et al., 2021). Moreover, the springiness and chewiness of fries and the hardness of the outer crust of the fries were improved (Leong et al., 2022; Zhang et al., 2021). The imposition of high-voltage pulses could also cause electroporation on the starch surface, which altered the gelatinization and pasting properties, retrogradation properties, and in-vitro digestibility. Potato tubers pretreated by PEF at an electric field strength of 1.9 kV/cm showed a lower percentage of total starch hydrolysis (Leong et al., 2022).

Overall, PEF could induce damage to the semicrystalline structure, granular intact, and granular size distribution of starches by disturbing the intermolecular hydrogen bonding (H–H) and the formation of double helices within amylopectin moieties (Giteru et al., 2018; Zhu, 2018). These attributes further determine the functional properties such as gelatinization temperature, gelatinization enthalpy, swelling power, pasting viscosity, water solubility, mechanical properties, and digestibility (Leong et al., 2022; Walter, 1998). Therefore, various potential applications of PEF-modified starch and PEF in improving the quality of starchy food have been proposed (Figure 5.6). Notably, these applications were almost conducted at the laboratory scale, and tremendous efforts still need to be made to achieve the industrial applications.

REFERENCE, BIBLIOGRAPHY OR WORKS CITED

Alalaiwe, A., Fayed, M. H., Alshahrani, S. M., Alsulays, B. B., Alshetaili, A. S., Tawfeek, H. M., & Khafagy, E. S. 2019. Application of design of experiment approach for investigating the effect of partially pre-gelatinized starch on critical quality attributes of rapid orally disintegrating tablets. *Journal of Drug Delivery Science and Technology* 49: 227–234.

Almeida, R. L. J., Santos, N. C., Padilha, C. E., de Almeida Mota, M. M., de Alcântara Silva, V. M., André, A. M. M. C. N., & dos Santos, E. S. 2022. Application of pulsed electric field and drying temperature response on the thermodynamic and thermal properties of red rice starch (*Oryza sativa* L.). *Journal of Food Process Engineering* 45 (2): e13947.

An, H. J., & King, J. M. 2010. Pasting properties of ohmically heated rice starch and rice flours. *Journal of Food Science* 71 (7): C437–C441.

Arshad, R. N., Abdul-Malek, Z., Munir, A., Buntat, Z., Ahmad, M. H., Jusoh, Y. M. M., Aadil, R. M. 2020. Electrical systems for pulsed electric field applications in the food industry: An engineering perspective. *Trends in Food Science & Technology* 104: 1–13.

Bai, T. G., Zhang, L., Qian, J.-Y., Jiang, W., Wu, M., Rao, S. Q., Wu, C. 2021. Pulsed electric field pretreatment modifying digestion, texture, structure and flavor of rice. *LWT* 138: 110650.

BeMiller, J. N. 2016. Starch, modification. Reference Module in Food Science. London: Elsevier Academic Press.

Hoover, R. 1998. Starch-lipid interactions. In R. H. Walter (Ed.). *Polysaccharide Association Structures in Food*, New York: CRC Press.

Castro, L. M. G., Alexandre, E. M. C., Saraiva, J. A., & Pintado, M. 2023. Starch extraction and modification by pulsed electric fields. *Food Reviews International* 39 (4): 2161–2182.

Cha, Y. 2012. Effect of ohmic heating at subgelatinization temperatures on thermal-property of potato starch. *The Korean Journal of Food and Nutrition 25*(4): 1068–1074.

Cha, Y. H. 2014. Effect of ohmic heating on thermal and water holding property of starches. *The Korean Journal of Food and Nutrition* 27: 112–119.

Chen, B. R., Wen, Q. H., Zeng, X. A., Abdul, R., Roobab, U., & Xu, F. Y. 2021. Pulsed electric field assisted modification of octenyl succinylated potato starch and its influence on pasting properties. *Carbohydrate Polymers* 254: 117294.

Fonseca, L. M., Halal, S. L. M. E., Dias, A. R. G., & Zavareze, E. d. R. 2021. Physical modification of starch by heat-moisture treatment and annealing and their applications: A review. *Carbohydrate Polymers* 274: 118665.

Fried, S. D., Bagchi, S., & Boxer, S. G. 2014. Extreme electric fields power catalysis in the active site of ketosteroid isomerase. *Science* 346 (6216): 1510–1514.

Giteru, S. G., Oey, I., & Ali, M. A. 2018. Feasibility of using pulsed electric fields to modify biomacromolecules: A review. *Trends in Food Science & Technology* 72: 91–113.

Goullieux, A., & Pain, J. P. 2014. Chapter 22-ohmic heating. In D. W. Sun (Eds.), *Emerging technologies for food processing (Section Edition)* (pp. 399–426). San Diego: Academic Press.

Grgić, I., Ačkar, Đ., Barišić, V., Vlainić, M., Knežević, N., & Medverec Knežević, Z. 2019. Nonthermal methods for starch modification: A review. *Journal of Food Processing and Preservation* 43 (12): e14242.

Gulzar, S., Narciso, J. O., Elez-Martínez, P., Martín-Belloso, O., & Soliva-Fortuny, R. 2023. Recent developments in the application of novel technologies for the modification of starch in light of 3D food printing. *Current Opinion in Food Science* 52:101067.

Han, Z., Zeng, X. A., Fu, N., Yu, S. J., Chen, X. D., & Kennedy, J. F. 2012. Effects of pulsed electric field treatments on some properties of tapioca starch. Carbohydrate Polymers 89 (4): 1012–1017.

Han, Z., Zeng, X. A., Yu, S. J., Zhang, B. S., & Chen, X. D. 2009a. Effects of pulsed electric fields (PEF) treatment on physicochemical properties of potato starch. *Innovative Food Science & Emerging Technologies* 10 (4): 481–485.

Han, Z., Zeng, X. A., Zhang, B. S., & Yu, S. J. 2009b. Effects of pulsed electric fields (PEF) treatment on the properties of corn starch. *Journal of Food Engineering* 93 (3): 318–323.

Hoover, R. 1998. Starch-lipid interactions. In R. H. Walter (Ed.). *Polysaccharide association structures in food*. New York: CRC Press.

Hong, J., Chen, R., Zeng, X. A., & Han, Z. 2016. Effect of pulsed electric fields assisted acetylation on morphological, structural and functional characteristics of potato starch. *Food Chemistry* 192: 15–24.

Icier, F. 2012. Chapter 11-Ohmic heating of fluid foods. In P. J. Cullen, B. K. Tiwari & V. P. Valdramidis (Eds.), *Novel thermal and non-thermal technologies for fluid foods* (pp. 305–367). San Diego: Academic Press.

Ignat, A., Manzocco, L., Brunton, N. P., Nicoli, M. C., & Lyng, J. G. 2015. The effect of pulsed electric field pre-treatments prior to deep-fat frying on quality aspects of potato fries. *Innovative Food Science & Emerging Technologies* 29: 65–69.

Jaeger, H., Roth, A., Toepfl, S., Holzhauser, T., Engel, K.-H., Knorr, D., Steinberg, P. 2016. Opinion on the use of ohmic heating for the treatment of foods. *Trends in Food Science & Technology* 55: 84–97.

Janositz, A., Noack, A. K., & Knorr, D. 2011. Pulsed electric fields and their impact on the diffusion characteristics of potato slices. *LWT* 44 (9): 1939–1945.

Jayakody, L., & Hoover, R. 2008. Effect of annealing on the molecular structure and physicochemical properties of starches from different botanical origins - A review. *Carbohydrate Polymers* 74 (3): 691–703.

Kanjanapongkul, K. 2017. Rice cooking using ohmic heating: Determination of electrical conductivity, water diffusion and cooking energy. *Journal of Food Engineering* 192: 1–10.

Karapantsios, T. D., Sakonidou, E. P., & Raphaelides, S. N. 2010. Electrical conductance study of fluid motion and heat transport during starch gelatinization. *Journal of Food Science* 65 (1): 144–150.

Kaur, N., & Singh, A. K. 2015. Ohmic heating: Concept and applications - A review. *Critical Reviews in Food Science & Nutrition* 56 (14): 2338–2351.

Knirsch, M. C., Santos, C. A. d., Vicente, A. A. M. d. O. S., & Penna, T. C. V. 2010. Ohmic heating - A review. *Trends in Food Science & Technology* 21 (9): 437–441.

Leong, S. Y., Roberts, R., Hu, Z., Bremer, P., Silcock, P., Toepfl, S., & Oey, I. 2022. Texture and *in* vitro starch digestion kinetics of French fries produced from potatoes (Solanum tuberosum L.) pre-treated with pulsed electric fields. Applied Food Research 2 (2): 100194.

Li, D., Jiang, L., Tao, Y., Yang, N., & Han, Y. 2021. Enhancement of efficient and selective hydrolysis of maize starch via induced electric field. *LWT* 143: 111190.

Li, F. D., Li, L. T., Li, Z., & Tatsumi, E. 2004. Determination of starch gelatinization temperature by ohmic heating. *Journal of Food Engineering* 62 (2): 113–120.

Li, D., Liu, R., Tao, Y., Shi, Y., Wang, P., & Han, Y. 2023. Enhancement of the carboxymethylation of corn starch via induced electric field. *Carbohydrate Polymers* 319: 121137.

Li, D., Luo, X., Tao, Y., Wang, P., Yang, N., Xu, E., & Han, Y. 2022. Intensifying the moderate electric field-induced modification of maize starch by 1-butyl-3-methylimidazolium chloride. *Carbohydrate Polymers* 292: 119654.

Li, D., Yang, N., Jin, Y., Zhou, Y., Xie, Z., Jin, Z., & Xu, X. 2016. Changes in crystal structure and physicochemical properties of potato starch treated by induced electric field. *Carbohydrate Polymers* 153: 535–541.

Li, D., Zhang, Y., Yang, N., Jin, Z., & Xu, X. 2018. Impact of electrical conductivity on acid hydrolysis of guar gum under induced electric field. *Food Chemistry* 259: 157–165.

Li, D. D., Yang, N., Tao, Y., Xu, E. B., Jin, Z. Y., Han, Y. B., & Xu, X. M. 2020. Induced electric field intensification of acid hydrolysis of polysaccharides: Roles of thermal and non-thermal effects. *Food Hydrocolloids* 101: 105484.

Li, Q., Wu, Q. Y., Jiang, W., Qian, J. Y., Zhang, L., Wu, M. G., Wu, C. S. 2019. Effect of pulsed electric field on structural properties and digestibility of starches with different crystalline type in solid state. *Carbohydrate Polymers* 207: 362–370.

Li, Y., Wang, J. H., Han, Y., Yue, F. H., Zeng, X. A., Chen, B. R., Han, Z. 2023a. The effects of pulsed electric fields treatment on the structure and physicochemical properties of dialdehyde starch. *Food Chemistry* 408: 135231.

Li, Y., Wang, J. H., Wang, E. C., Tang, Z. S., Han, Y., Luo, X. E., Han, Z. 2023b. The microstructure and thermal properties of pulsed electric field pretreated oxidized starch. *International Journal of Biological Macromolecules* 235: 123721.

Liu, X., & Hong Y. 2018. Chapter 4-Pre-gelatinized Modification of Starch. In Z. Sui & X. Kong (Eds). *Physical modifications of starch* (pp. 51–60). Singapore: Springer.

Luo, X. E., Wang, R. Y., Wang, J. H., Li, Y., Luo, H. N., Zeng, X. A., Han, Z. 2023. Combining pulsed electric field and cross-linking to enhance the structural and physicochemical properties of corn porous starch. *Food Chemistry* 418: 135971.

Maniglia, B. C., Pataro, G., Ferrari, G., Augusto, D., Le-Bail, P., Le-Bail, A. 2021. Pulsed electric fields (PEF) treatment to enhance starch 3D printing application: Effect on structure, properties, and functionality of wheat and cassava starches. *Innovative Food Science and Emerging Technologies* 68: 102602.

Martínez-Bustos, F., López-Soto, M., Zazueta-Morales, J. J., & Morales-Sánchez, E. 2005. Preparation and properties of pre-gelatinized cassava (*Manihot esculenta. Crantz*) and Jícama (*Pachyrhizus erosus*) starches using ohmic heating. *Agrociencia* 39 (3): 275–283.

Mimura, K., Kanada, K., Uchida, S., Yamada, M., & Namiki, N. 2011. Formulation study for orally disintegrating tablet using partly pregelatinized starch binder. *Chemical & Pharmaceutical Bulletin (Tokyo)* 59 (8): 959–964.

Pinto, J., Silva, V. L. M., Silva, A. M. G., Silva, A. M. S., Costa, J. C. S., Santos, L. M. N. B. F.,... Teixeira, J. A. C. 2013. Ohmic heating as a new efficient process for organic synthesis in water. *Green Chemistry* 15 (4): 970–975.

Pryor, R. W. 2013. Inductive conductivity measurement of seawater. *Proceedings of the COMSOL Conference* in Boston.

Qiu, S., Abbaspourrad, A., & Padilla-Zakour, O. 2021. Changes in the glutinous rice grain and physicochemical properties of its starch upon moderate treatment with pulsed electric field. *Foods* 10 (2): 395.

Qiu, S., Abbaspourrad, A., & Padilla-Zakour, O. I. 2022. Prevention of the retrogradation of glutinous rice gel and sweetened glutinous rice cake utilizing pulsed electric field during refrigerated storage. *Foods* 11 (9): 1306.

Redondo, O., Prolongo, S., Campo, M., Sbarufatti, C., Giglio, M. 2018. Anti-icing and de-icing coatings based Joule's heating of graphene nanoplatelets. *Composites Science and Technology* 164: 65–73.

Sakr, M., & Liu, S. 2014. A comprehensive review on applications of ohmic heating (OH). *Renewable and Sustainable Energy Reviews* 39: 262–269.

Singh, S., Sharanagat, V. S., Desai, S., Kumar, K., Upadhyay, S., & Chakraborty, G. J. 2023. Interaction of pulsed electric modified elephant foot yam (Amorphophallus paeoniifolius) starch and polyvinyl alcohol to enhance the mechanical and barrier properties of film. *Starch-Stárke*: 2200229.

Sung, J., Park, D., Park, B., Choi, H., & Jhon, M. J. B. 2005. Phosphorylation of potato starch and its electrorheological suspension. *Biomacromolecules* 6 (4): 2182–2188.

Svihus, B., Uhlen, A. K., & Harstad, O. M. 2005. Effect of starch granule structure, associated components and processing on nutritive value of cereal starch: A review. *Animal Feed Science and Technology* 122 (3): 303–320.

Tester, R. F., & Debon, S. J. J. 2000. Annealing of starch-a review. *International Journal of Biological Macromolecules* 27 (1): 1–12.

Varella, V., Concone, B., Senise, J., & Doin, P. 1984. Continuous enzymatic cooking and liquefaction of starch using the technique of direct resistive heating. *Biotechnology and Bioengineering* 26 (7): 654–657.

Wang, B. X., & Zhao, X. P. 2002. Electrorheological effect coordinated by kaolinite-carboxymethyl starch hybrid materials. *Journal of Materials Chemistry* 12 (10): 2869–2871.

Wang, W. C., & Sastry, S. K. 1997. Starch gelatinization in ohmic heating. *Journal of Food Engineering* 34 (3): 225–242.

Wu, C., Wu, Q. Y., Wu, M. G., Jiang, W., Qian, J. Y., Rao, S. Q., Zhang, C. 2019. Effect of pulsed electric field on properties and multi-scale structure of japonica rice starch. *LWT* 116: 108515.

Yang, N., Jin, Y., Li, D., Jin, Z., & Xu, X. 2017. A reconfigurable fluidic reactor for intensification of hydrolysis at mild conditions. *Chemical Engineering Journal* 313: 599–609.

Zavareze, E. d . R., & Dias, A. R. G. 2011. Impact of heat-moisture treatment and annealing in starches: A review. *Carbohydrate Polymers* 83 (2): 317–328.

Zeng, F., Gao, Q. Y., Han, Z., Zeng, X. A., & Yu, S. J. 2016. Structural properties and digestibility of pulsed electric field treated waxy rice starch. *Food Chemistry* 194: 1313–1319.

Zhang, C., Lyu, X., Arshad, R. N., Aadil, R. M., Tong, Y., Zhao, W., & Yang, R. 2022. Pulsed electric field as a promising technology for solid foods processing: A review. *Food Chemistry* 403: 134367.

Zhang, C., Lyu, X., Zhao, W., Yan, W., Wang, M., Rei, N. K., Yang, R. 2021. Effects of combined pulsed electric field and blanching pretreatment on the physiochemical properties of French fries. *Innovative Food Science and Emerging Technologies* 67: 102561.

Zhu, F. 2018. Modifications of starch by electric field based techniques. *Trends in Food Science & Technology* 75: 158–169.

CHAPTER 6

Microwave Treatment of Starch and Starchy Materials

Xuejiao Zhang and Yongbin Han

6.1 INTRODUCTION

Microwaves are electromagnetic waves with centimeter-long wavelengths that are often used at frequencies between 300 MHz and 300 GHz. Common frequencies (especially used in the food processing industry) include 2.45 and 0.915 GHz, with corresponding wavelengths of 12.25 and 32.80 cm, respectively (Fan et al., 2017a). As a new physical technology, microwaves have gradually attracted great interest in recent years due to their high efficiency, simple operation, and no by-products (Samson et al., 2021). It has the ability to alter the structural properties of materials during processing and is therefore commonly used to treat and modify the structural properties of starches (Guo et al., 2017). Since the 1990s, a large number of studies on the interaction of microwaves with starch have appeared, and some relevant studies are presented in the form of a timeline, as shown in Figure 6.1.

Current studies mainly investigated the effects of microwaves with different powers and durations on the physicochemical and functional properties as well as the structure of different types of starch. In general, microwaves can cause the rearrangement of intramolecular structures, which alters the water absorption, solubilization, swelling, pasting, dehydration shrinkage,

DOI: 10.1201/9781003493594-6

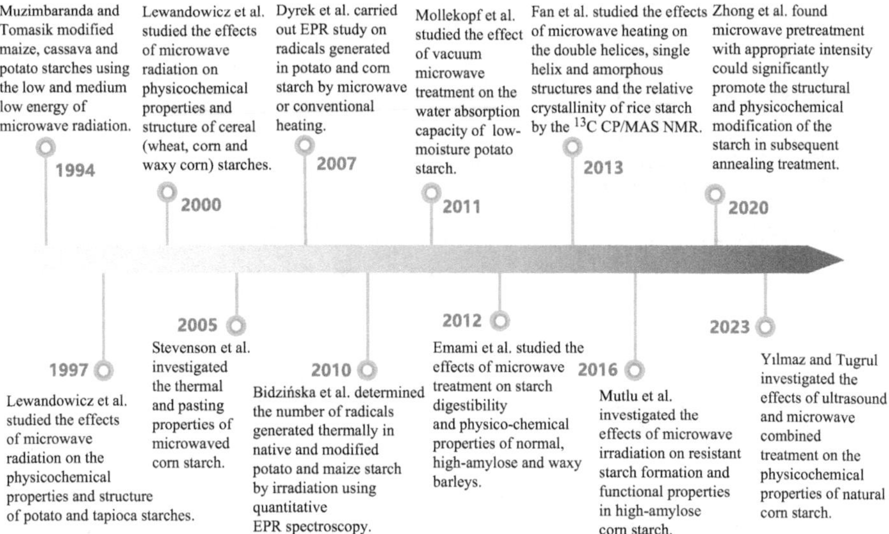

FIGURE 6.1 Timeline for important events in modifying starch via microwave radiation.

and rheology of starch and in turn affects its functionalities (Zeng et al., 2016). In this chapter, we described the mechanism of action of microwaves and the structural changes in starch caused by the microwave response after microwave radiation. Also, the possible applications of microwave-treated starch (MTS) are summarized. The aim of this work is to lay the theoretical foundation for future research into the creation and promotion of microwave-modified starch (MMS) and starchy materials.

6.2 PRINCIPLE OF MICROWAVE ACTION

The principle of microwave acting on matter can be divided into two types: microwave photon irradiation and microwave dielectric heating (Brasoveanu & Nemtanu, 2014), as shown in Figure 6.2. Microwave photon irradiation directly acts on the receptor through the energy of electromagnetic photons. The microwave photon energy of 2.45 GHz cannot destroy the hydrogen bond, which makes it more difficult to affect the covalent bond in the molecule. Thus, it is evident that direct absorption of microwave radiation to stimulate chemical reactions is not possible (Fan et al., 2012a). Microwave dielectric heating is a kind of heating technology that absorbs microwave energy through electromagnetic induction material and converts it into heat at the same time (El Khaled et al., 2018). The majority of the more sophisticated chemical reactions in

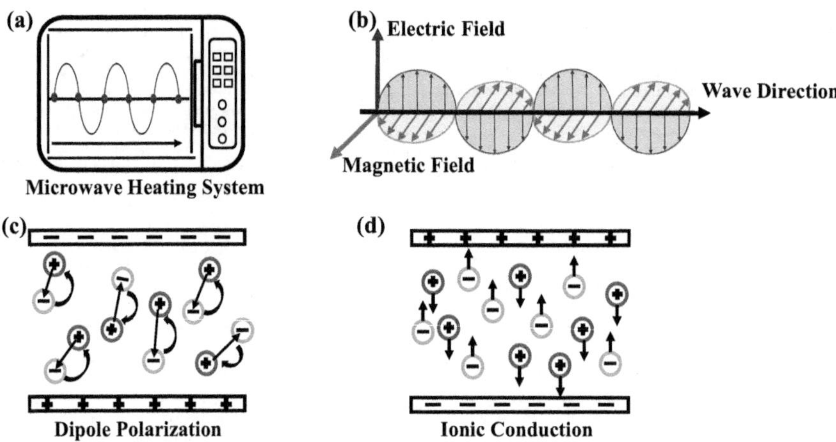

FIGURE 6.2 The principle of microwave action: (a) A microwave heating system, (b) the electromagnetic field in the microwave system, (c) the dipole polarization mechanism, and (d) the ionic conduction mechanism.

which microwaves engage are based on the microwave dielectric heating mechanism (Lewicka et al., 2015).

In the microwave electromagnetic composite field, the electric field serves as the primary source of the thermal effect, which has two mechanisms: dipole polarization and ionic conduction (Sun et al., 2016). The dipoles are rearranged due to the phase difference between the electric field direction and the induction dipole, which causes the molecular collision and friction process resulting in the conversion of microwave energy into heat. This mechanism is called dipole polarization (Singh et al., 2015). Microwaves cause ions in solution to oscillate in response to changes in the electric field and collide with surrounding atoms and molecules to produce microscopic displacements leading to heat, which is the ionic conduction mechanism (Clark et al., 2000). Moreover, comparing the two mechanisms, the thermal effect produced by the ionic conduction process is stronger (Fan et al., 2012b).

There are many factors that affect the heat distribution in the microwave heating process, the most important of which are the dielectric property and penetration depth of materials (Mudgett, 1982). The dielectric property means that the bound charge in the material shows the property of saving and losing electromagnetic energy under the action of an electric field (Fan et al., 2015). Furthermore, the dielectric property is mainly characterized by the relative dielectric constant (ε'), relative dielectric loss factor (ε''), and dielectric loss angular tangent ($\tan \delta\varepsilon$) (Magee et al., 2013).

ε' is used to describe the nature that matter molecules can be polarized. ε'' is used to describe the ability of matter molecules to convert electromagnetic energy into thermal energy. tan$\delta\varepsilon$ reflects the ability of matter to absorb electromagnetic energy into thermal energy at a specified temperature and power. The higher tan$\delta\varepsilon$ is, the more effective microwave energy is absorbed and the faster the temperature rises; the net value of tan$\delta\varepsilon$ can be calculated as $\varepsilon''/\varepsilon'$ (Tao et al., 2020). The penetration depth is controlled by the dielectric constant of the material, and thus the penetration depth mainly depends on the frequency of the microwave and the dielectric property of the material itself (Zhu & Guo, 2017).

Starch is an important macromolecular substance, its dielectric property often reflects the microwave response characteristic and determines the microwave penetration depth in products (Fan et al., 2017b). The dielectric constant and dielectric loss factor of starch mainly depend on water content (Lee et al., 2019), temperature (Motwani et al., 2007), starch source (Ndife et al., 1998), and microwave frequency (Nelson & Trabelsi, 2006). Starch absorbs electromagnetic energy and transforms it into kinetic energy due to its dielectric heating capability, and its molecular conformation, the vibrational strength of its chemical bonds, and the intensity of the accompanying infrared absorption peaks may alter as a result (Xu et al., 2019). Meanwhile, starch molecules frictionally collide with each other to generate thermal energy, thus increasing the temperature (Li et al., 2019a). Starch contains a large amount of –OH internally, which specifically combines with H_2O during conventional heat treatment leading to pasting. During microwave heating, the high dielectric property of H_2O accelerated the pasting process of starch, significantly affecting the degree of reaction of the whole system (Shen et al., 2017).

Due to the special mechanism of microwave heating, it has the following characteristics: (1) holistic and fast heating. Microwave energy acts on all parts of the heated material at the same time. When starch is placed in a microwave electromagnetic field, energy is absorbed irreversibly, resulting in rapid volumetric heating. (2) Selective heating and low energy loss. The dielectric properties of different materials are not the same, and their ability to absorb microwaves to convert them into heat is also different. Microwaves can be absorbed by materials with a high dielectric loss factor, such as water, while passing through materials with a low dielectric loss factor, such as ceramics, resulting in selective heating (Lin et al., 2022). Therefore, the effectiveness of microwave heating of starch and starchy materials can be improved by appropriately raising the moisture content and using ceramics as the container material. (3) The heating process can be controlled. The thermal inertia of microwave heating is small, and it can be stopped immediately after the input of microwave heating is stopped. Microwave heating of starch can be started as soon as the parameters are set. And once the heating is stopped, there is basically no residual heat. (4) Non-thermal effect. At present, whether there is a non-thermal effect of microwave has been debated, but microwave can significantly increase or reduce the reaction rate of chemicals. For example, the non-thermal effect of microwave radiation reduces the apparent activation energy of the esterification reaction of tapioca starch (Shi et al., 2016).

6.3 STRUCTURAL CHANGE OF MICROWAVE-TREATED STARCH (MTS)

Starch is mainly composed of two kinds of α-glucans (amylose and amylopectin), accounting for 98%–99% of the dry weight (Tester et al., 2004). Although glucose is the basic constituent unit of starch, its structure is very complex (Oyeyinka & Oyeyinka, 2018). Among them, the multi-scale structure model of starch proposed by Corre et al. (2010) has been recognized by most studies. According to the multi-scale structure model of starch, starch granules contain a variety of characteristic structures of different scales formed by intramolecular and intermolecular hydrogen bonding interactions (for details, see Chapter 1) (Huang et al., 2014). At present, scanning electron microscopy (SEM) (Utrilla-Coello et al., 2013), X-ray diffraction (Zhong et al., 2020), ^{13}C cross-polarization/magic angle spinning nuclear magnetic resonance (^{13}C CP/MAS NMR) (Xu et al., 2019), Fourier transform infrared spectroscopy (FTIR) (Zhang et al., 2019) and Raman spectroscopy (Raman) (Fan et al., 2012c) are mainly used to study the changes in starch multi-scale structure after microwave radiation. And in Sections 6.3.1–6.3.4, the structural changes of starch caused by microwave treatment were discussed.

6.3.1 Granule Morphology

The effect of microwave radiation on the granule morphology in the multi-scale structure of starch is shown in Figure 6.3. The effect of microwave on the morphology of starch granules was closely related to the treatment time. Pits and voids were discovered in microwave-treated corn starch granules by Yılmaz and Tugrul (2023). When the microwave time increased, the granules deformed and broke, eventually losing their integrity. Oyeyinka et al. (2019) reported that the degree of distortion on the surface of *Bambara*

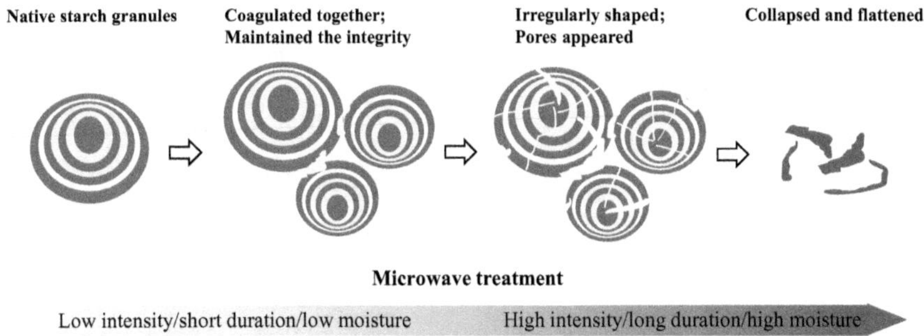

FIGURE 6.3 Effect of microwave radiation on the granule morphology of starch.

groundnut starch granules increased with increasing microwave time. After 10 seconds of microwave treatment, some of the granules coagulated together, but the integrity of the granules was maintained. After 30 seconds, most of the granules of starch were irregularly shaped and pores could be seen on a few granules. After 60 seconds, the structure of most of the starch granules collapsed and the granules flattened. Kumar et al. (2020) also reported the morphological response of potato starch granules to time-varying microwave heating. SEM micrographs showed unaffected starch granules at 1 minute of treatment while fissures and indentations developed at 3 and 5 minutes. In addition, the effect of microwave on the morphology of starch granules was related to the moisture content of starch. Under the same microwave treatment conditions, Solaesa et al. (2022) found that the polygonal structure of rice starch granules with 8% water content disappeared and small pores appeared, whereas starch granules with 30% water content showed rounded aggregates. Li et al. (2019b) also found that starch with 30% moisture content kept its granular morphology basically unchanged and only some cracks appeared on the surface, while starch with 40% moisture content aggregated together to form larger clusters. When the moisture content further rose to 45%, the starch granules were completely deformed by the microwave and even became gelatinous.

6.3.2 Lamellar Structure

The effect of microwave radiation on the lamellar structure in the multi-scale structure of starch depends on the starch origins and microwave conditions, which is shown in Figure 6.4. Tian et al. (2023) found that microwave treatment increased the thickness of the amorphous layer and decreased the thickness of the crystalline layer of maize starch, while the

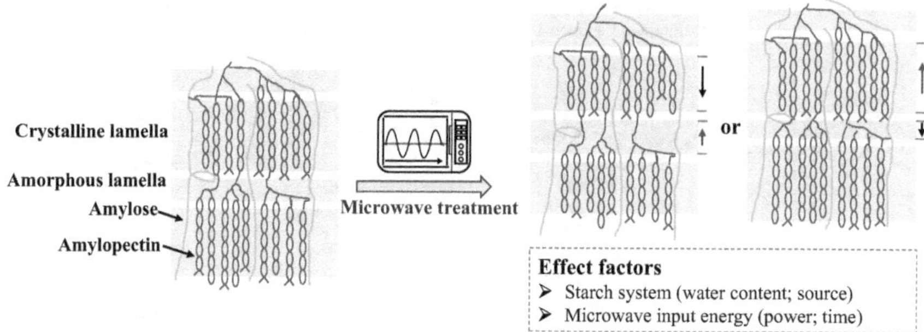

FIGURE 6.4 Effect of microwave radiation on the lamellar structure of starch.

Bragg sheet repeat distance and the long-period distance remained almost unchanged. Moreover, at 800 W, the lamellar structural changes in maize starch were more significant than those at 400 W, but similar to those at 1200 W. Therefore, the effect of microwaves on the starch lamellar structure is related to microwave power but not linearly. However, the intensity of the scattering peaks of the SAXS curves of rice starch with high moisture content decreased with increasing temperature, which corresponded to a smaller thickness of the amorphous layer and an increase in the thickness of the crystalline layer (Fan et al., 2013b). Yan et al. (2019) also discovered similar trends of amorphous and crystalline layers of starch in microwave treatments of rice starch with low moisture content, which suggested that the microwave action made the amorphous layer of starch more compact.

6.3.3 Crystalline Structure

The effect of microwave radiation on the crystalline structure in the multi-scale structure of starch is shown in Figure 6.5. Sun et al. (2023) found that the relative crystallinity of microwave-treated lotus root starch was lower than that of the native starch. Yang et al. (2017) discovered that after 5, 10, and 20 minutes of microwave action, the relative crystallinity of waxy maize starch dropped to 10.47%, 6.42%, and 2.91%, respectively. This could be caused by that microwaves unspin the double-helix structure of waxy maize starch, resulting in a drop in relative crystallinity. In contrast, Xia et al. (2018) in their study found that the relative crystallinity of potato starch increased instead after microwave treatment. These two completely different phenomena may be attributed to the differences in microwave equipment parameters and starch crystallization types. During microwave heating, moisture leaves the starch granules due to the high temperature, and the increase in temperature or the reorganization of starch

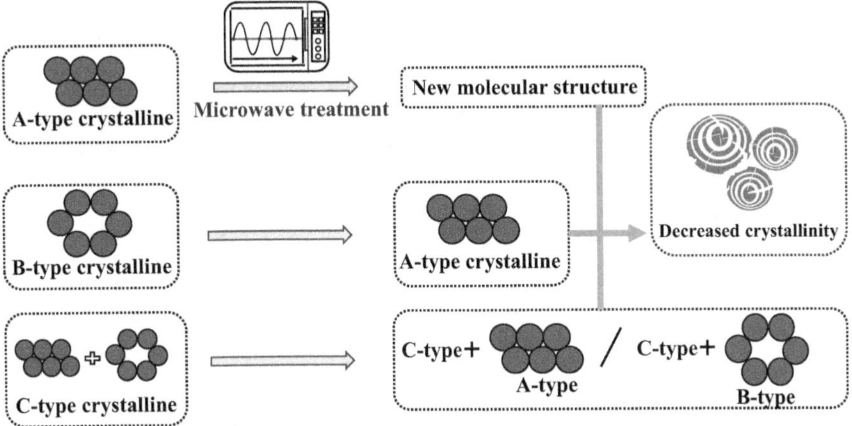

FIGURE 6.5 Effect of microwave radiation on the crystalline structure of starch.

chains may have different effects on the crystallinity of starch. In addition, the microwave action of larger power can even change the crystalline structure of starch. Nawaz et al. (2018) showed that the crystalline form of lotus starch after microwave treatment was transformed from C-type to a mixed form of C-type and A-type or a mixed form of C-type and B-type. Similarly, the crystal structure of potato starch changed from B-type to A-type crystallization after microwave treatment (Xu et al., 2019). Li et al. (2020) found that microwave-treated rice starch underwent starch chain reassembly during digestion, resulting in the transformation of the starch from an A-crystalline form to a new molecular structure.

6.3.4 Molecular Structure

The effect of microwave radiation on the molecular structure in the multi-scale structure of starch is shown in Figure 6.6. Bond breaks that result in shorter chains and glucose ring vibrations that alter the configuration and conformation of chain segments are the main causes of changes in the molecular chain structure of starch (Li & Zhu, 2017). Microwaves act on the chemical bonds and chemical groups in starch, thus affecting the vibrational intensity, the absorption peak intensity, and the peak widths of the groups in the infrared spectrum. The polar groups in starch molecules may change their molecular conformation, which leads to a change in vibration intensity, and the corresponding intensity of infrared absorption peaks may also be changed (Chen et al., 2021). Yang et al. (2017) found that the number of branched amylose short chains (A chains) in waxy maize starch after microwave radiation was less, and the proportion of short B1 and long

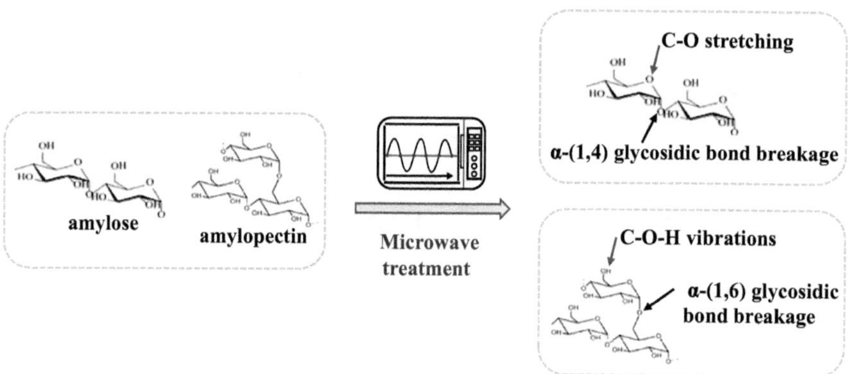

FIGURE 6.6 Effect of microwave radiation on the molecular structure of starch.

B2 and B3 was higher. Moreover, ^{1}H NMR data showed that the α-(1,6) glycosidic bond was more easily destroyed than the α-(1,4) glycosidic bond during the microwave treatment. Kizil et al. (2002) reported that the FTIR spectroscopy of corn starch changed after microwave radiation, including C–O stretching (1040 and 1150 cm^{-1}), C–O–H vibrations (950 cm^{-1}). In addition, the relative molecular mass and mean square radius of gyration of lotus starch decreased with the increase of microwave power, which also proved that microwaves could degrade starch chains (Zeng et al., 2016).

Overall, microwaves act on the molecular chain structure of starch, which in turn alters its crystalline structure, lamellar structure, and granular morphology. There are numerous studies on the impact of microwaves on the multi-scale structure of starch, but the majority of them focused on the starch system itself (moisture content and source) as well as the microwave input energy (microwave power and time). These investigations mostly ignored the interactions between the microwave and the starch system. It is also challenging to explain the entire process based on a single property because the characteristics of starch systems, such as moisture content and density, may influence their reaction to microwave heating in a variety of ways. Therefore, additional in-depth research is still urgently required to comprehend how microwaves affect the multi-scale structure of starch.

6.4 MTS APPLICATIONS

In recent years, microwave technology has become increasingly popular in the field of modifying starch and starchy materials due to its advantages of fast and uniform heating, simple operation, and environmental friendliness. In general, microwave radiation causes molecular rearrangement within the starch granules dependent on the dielectric characteristics of starch, thus altering the structure of starch including granular morphology, lamellar structure, crystalline structure, and molecular structure. The effect of microwave on the multi-scale structure of starch will directly cause changes in its functional properties, resulting in a physically modified starch structure with excellent processing characteristics. And MTS are already used in food and non-food fields (such as medicine and packaging) these years. Furthermore, in the 3D printing field, microwave technology can significantly improve the rheological and mechanical properties of starch-based materials for 3D printing, making them more suitable for 3D printing (details in Chapter 10). For example, Xu et al. (2020) demonstrated that wheat starch under microwave treatment led to optimal support stability, line consistency, and high retention of 3D-printed products; good printing accuracy was achieved due to microwave treatment induced-partial pasting of wheat starch (Sun et al., 2020). However, despite the great potential of microwave technology for the modification of starch and starchy materials, there is still a long way to go before its industrial applications can be realized.

6.4.1 Medicine and Packaging as Non-Food Use

The effect of microwave on the multi-scale structure of starch will directly alter its physicochemical properties such as solubility, gelatinization, retrogradation, and digestibility, thus ultimately resulting in the formation of modified starch with excellent processing characteristics. Various applications of MMS or starch-based materials have been reported (Figure 6.7). In the pharmaceutical field, Luu et al. (2015) found that microwave-modified rice starch controlled the drug release in simulated gastric fluid, suggesting that it could be a promising material for sustained-release dosage forms. Sriamornsak et al. (2010) discovered that MMS as a hydrophilic matrix excipient of sustained-release tablets exhibited superior properties compared to native starch. In the packaging field, cassava starch-based biodegradable films were prepared using a microwave-assisted method by Souza et al. (2012). However, Zhu et al. (2016) found that the multi-scale structural changes of MMS led to a weakening of its barrier properties, which facilitated the migration of plasticizers. This study provided fundamental data for rationally adjusting the structural characteristics of starch-based films to control the migration of chemicals, thus making it possible to design more rational starch-based films for packaging.

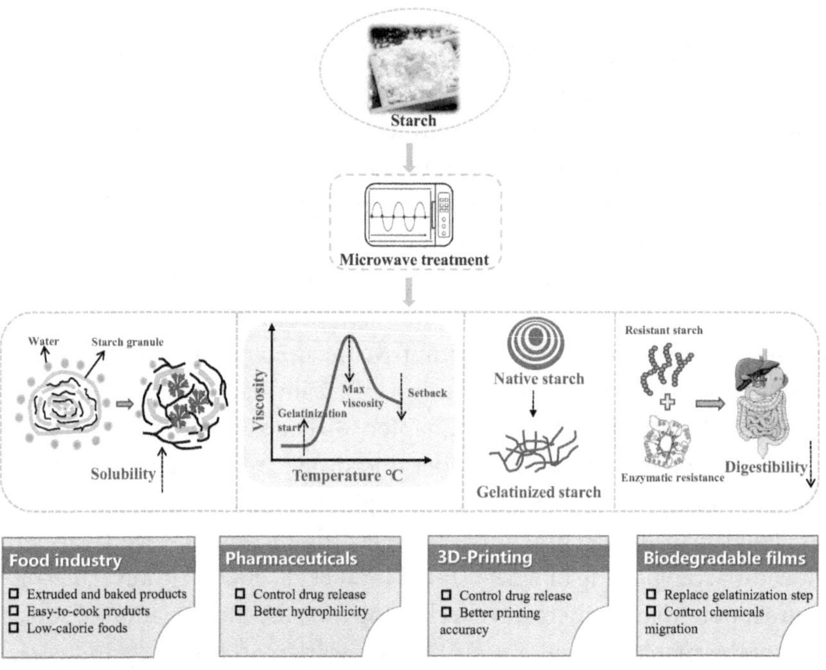

FIGURE 6.7 Effect of microwave radiation on the physicochemical properties of starch and its potential applications.

6.4.2 MTS as Potential Food Use

There seems no food application for MMS in business (except for MTS or starchy foods like cereals) currently, but it is believed that using microwave heating alone for preparing modified starch or in combination with other technologies to change starches has significant potential for food applications. For example, MMS can be used to reduce the aging tendency of bread, which is a result of its reduced reflux viscosity (Tian et al., 2023). They can also be used in applications requiring high temperatures, such as extruded and baked products since microwaves improve the thermal stability of starches (Wu et al., 2022). The reduced digestibility of the starch induced by microwaves can be employed to produce low-calorie foods including those developed for the treatment of diabetes (Li et al., 2021). In addition, microwave heating leads to partial pasting of the starch, resulting in faster water absorption than unpasted pasta (Zhang et al., 2023). Thus, MMSs may be added to dough or other flour compositions to reduce the time it takes to cook the pasta. It should be noted that all of the preceding research simply demonstrates the possibility of MMS usage in food products. In future, researches should concentrate on the applications of MMS/MTS in model food systems.

Over the past few decades, researches related to microwave technology have mostly concentrated on the integration of control and sensing technology, as well as the development of shielding materials. The applications in other fields such as food and pharmaceuticals are still in their early stages, as evidenced by the fact that starchy raw materials and processed products cannot form a better match with the microwave treatment process, and the product quality often deteriorates after microwave processing or reheating. This is because that research on the impacts of microwaves on starchy materials is limited, and basic hypotheses on the effects of microwaves on the structure, function, and safety of starch and starchy materials have not yet been perfected. The aforementioned difficulties have hampered the development of microwave processing technology in food, pharmaceutical, and other industries. Therefore, the widespread use of MMS and starchy materials remains a great challenge.

REFERENCE, BIBLIOGRAPHY OR WORKS CITED

Bidzińska, E., Krystyna, D., & Elżbieta, W. 2010. Electron paramagnetic resonance study of thermally generated radicals in native and modified starches. *Current Topics in Biophysics* 33: 21–25.

Brasoveanu, M., & Nemtanu, M. R. 2014. Behaviour of starch exposed to microwave radiation treatment. *Starch-Stärke* 66: 3–14.

Clark, D. E., Folz, D. C., & West, J. K. 2000. Processing materials with microwave energy. *Materials Science and Engineering: A* 287 (2): 153–158.

Corre, D. L., Bras, J., & Dufresne, A. 2010. Starch nanoparticles: A review. *Biomacromolecules* 11: 1139–1153.

Chen, X. J., Liu, Y., Xu, Z. K., Zhang, C. C., Liu, X. X., Sui, Z. Q., & Corke, H. 2021. Microwave irradiation alters the rheological properties and molecular structure of hull-less barley starch. *Food Hydrocolloids* 120: 106821.

Dyrek, K., Bidzińska, E., Łabanowska, M., Fortuna, T., Przetaczek, I., & Pietrzyk, S. 2007. EPR study of radicals generated in starch by microwaves or by conventional heating. *Starch-Stärke* 59: 318–325.

El Khaled, D., Novas, N., Gazquez, J. A., & Manzano-Agugliaro, F. 2018. Microwave dielectric heating: Applications on metals processing. *Renewable & Sustainable Energy Reviews* 82: 2880–2892.

Emami, S., Perera, A., Meda, V., & Tyler, R. T. 2012. Effect of microwave treatment on starch digestibility and physico-chemical properties of three barley types. *Food and Bioprocess Technology* 5: 2266–2274.

Fan, D. M., Ma, S. Y., Wang, L. Y., Zhao, J. X., Zhang, H., & Chen, W. 2012a. Effect of microwave heating on optical and thermal properties of rice starch. *Starch-Stärke* 64: 740–744.

Fan, D. M., Li, C. X., Ma, W. R., Zhao, J. X., Zhang, H., & Chen, W. 2012b. A study of the power absorption and temperature distribution during microwave reheating of instant rice. *International Journal of Food Science & Technology* 47: 640–647.

Fan, D. M., Ma, W. R., Wang, L. Y., Huang, J. L., Zhao, J. X., Zhang, H., & Chen, W. 2012c. Determination of structural changes in microwaved rice starch using Fourier transform infrared and Raman spectroscopy. *Starch-Stärke* 64: 598–606.

Fan, D. M., Ma, W. R., Wang, L. Y., Huang, J. J., Zhang, F. M., Zhao, J. X., Zhao, H., & Chen, W. 2013a. Determining the effects of microwave heating on the ordered structures of rice starch by NMR. *Carbohydrate Polymers* 92: 1395–1401.

Fan, D. M., Wang, L. Y., Ma, S. Y., Ma, W. R., Liu, X. M., Huang, J. L., Zhao, J. X., Zhang, H., & Chen, W. 2013b. Structural variation of rice starch in response to temperature during microwave heating before gelatinization. *Carbohydrate Polymers* 92: 1249–1255.

Fan, D. M., Shen, H. J., Huang, L. L., Gao, Y. S., Lian, H. Z., Zhao, J. X., Zhang, H., & Chen, W. 2015. Microwave-absorbing properties of rice starch. *Polymers* 7: 1895–1904.

Fan, D. M., Wang, L. Y., Zhang, N. N., Xiong, L., Huang, L. L., Zhao, J. X., Wang, M. F., & Zhang, H. 2017a. Full-time response of starch subjected to microwave heating. *Scientific Reports* 7: 3967.

Fan, D. M., Gao, Y. S., Chen, Y. F., Wang, M. F., Gu, X. H., Wang, L. Y., Shen, H. J., Lian H. Z., Zhao, J. X., & Zhang, H. 2017b. Non-additive response of starch systems in different hydration states: A study of microwave-absorbing properties. *Innovative Food Science & Emerging Technologies* 44: 103–108.

Guo, Q. S., Sun, D. W., Cheng, J. H., Han, Z. 2017. Microwave processing techniques and their recent applications in the food industry. *Trends in Food Science & Technology* 67: 236–247.

Huang, J. R., Wei, N. G., Li, H. L., Liu, S. X., & Yang, D. Q. 2014. Outer shell, inner blocklets, and granule architecture of potato starch. *Carbohydrate Polymers* 103: 355–358.

Kizil, R., Irudayaraj, J., & Seetharaman, K. 2002. Characterization of irradiated starches by using FT-Raman and FTIR spectroscopy. *Journal of Agricultural and Food Chemistry* 50: 3912–3918.

Kumar, Y., Singh, L., Sharanagat, V. S., Patel, A., & Kumar, K. 2020. Effect of microwave treatment (low power and varying time) on potato starch: Microstructure, thermo-functional, pasting and rheological properties. *International Journal of Biological Macromolecules* 155: 27–35.

Lee, D. Y., Mo, C.Y., Lee, C. J., & Lee, S. H. 2019. Change in dielectric properties of sweet potato during microwave drying. *Food Science and Biotechnology* 28: 731–739.

Li, Y., Hu, A. J., Zheng, J., & Wang, X. Y. 2019a. Comparative studies on structure and physiochemical changes of millet starch under microwave and ultrasound at the same power. *International Journal of Biological Macromolecules* 141: 76–84.

Li, Y., Hu, A. J., Wang, X. Y., & Zheng, J. 2019b. Physicochemical and in vitro digestion of millet starch: Effect of moisture content in microwave. *International Journal of Biological Macromolecules* 134: 308–315.

Li, R., Dai, L. Y., Peng, H., Jiang, P., Liu, N. A., Zhang, D. J., Wang, C. Y., & Li, Z. J. 2021. Effects of microwave treatment on sorghum grains: Effects on the physicochemical properties and in vitro digestibility of starch. *Journal of Food Process Engineering* 44: e13804.

Li, N. N., Wang, L. L., Zhao, S. M., Qiao, D. L., Jia, C. H., Niu, M., Lin, Q. L., & Zhang, B. J. 2020. An insight into starch slowly digestible features enhanced by microwave treatment. *Food Hydrocolloids* 103: 105690.

Lin, X. H., Lyng, J., O'Donnell, C., & Sun, D. W. 2022. Effects of dielectric properties and microstructures on microwave-vacuum drying of mushroom (*Agaricus bisporus*) caps and stipes evaluated by non-destructive techniques. *Food Chemistry* 367: 130698.

Li, G. T., & Zhu, F. 2017. Molecular structure of quinoa starch. *Carbohydrate Polymers* 158: 124–132.

Lewandowicz, G., Fornal, J., & Walkowski, A. 1997. Effect of microwave radiation on physico-chemical properties and structure of potato and tapioca starches. *Carbohydrate Polymers* 34: 213–220.

Lewandowicz, G., Jankowski, T., & Fornal, J. 2000. Effect of microwave radiation on physico-chemical properties and structure of cereal starches. *Carbohydrate Polymers* 42: 193–199.

Lewicka, K., Siemion, P., & Kurcok, P. 2015. Chemical modifications of starch: Microwave effect. *International Journal of Polymer Science* 2015: 1–10.

Luu, T. D., Phan, N. H., Tran, T. T-D., Van Vo, T., & Tran, P. H-L. 2015. Use of microwave method for controlling drug release of modified sprouted rice starch. *In 5th International Conference on Biomedical Engineering in Vietnam.* Springer.

Magee, T. R. A., McMinn, W. A. M., Farrell, G., Topley, L., AI- Degs, Y. S., Walker, G. M., & Khraisheh, M. 2013. Moisture and temperature dependence of the dielectric properties of pharmaceutical powders. *Journal of Thermal Analysis and Calorimetry* 111: 2157–2164.

Mollekopf, N., Treppe, K., Fiala, P., & Dixit, O. 2011. Vacuum microwave treatment of potato starch and the resultant modification of properties. *Chemie Ingenieur Technik* 83: 262–272.

Motwani, T., Seetharaman, K., & Anantheswaran, R. C. 2007. Dielectric properties of starch slurries as infuenced by starch concentration and gelatinization. *Carbohydrate Polymers* 67: 73–79.

Mudgett, R. 1982. Electrical properties of foods in microwave processing. *Food Technology* 36: 109–115.

Mutlu, S., Kahraman, K., & Öztürk, S. 2016. Optimization of resistant starch formation from high amylose corn starch by microwave irradiation treatments and characterization of starch preparations. *International Journal of Biological Macromolecules* 95: 635–642

Muzimbaranda, C., & Tomasik, P. 1994. Microwaves in physical and chemical modifications of starch. *Starch-Stärke* 46(12): 469–474.

Nawaz, H., Shad, M. A., Saleem, S., Khan, M. U. A., Nishan, U., Rasheed, T., Bilal, M., & Iqbal, H. M. N. 2018. Characteristics of starch isolated from microwave heat treated lotus (*Nelumbo nucifera*) seed flour. *International Journal of Biological Macromolecules* 113: 219–226.

Ndife, M. K., Sumnu, G., & Bayindirli, L. 1998. Dielectric properties of six different species of starch at 2450 MHz. *Food Research International* 31: 43–52.

Nelson, S. O., & Trabelsi, S. 2006. Dielectric spectroscopy of wheat from 10 MHz to 1.8 GHz. *Measurement Science and Technology* 17: 2294–2298.

Oyeyinka, S. A., & Oyeyinka, A. T. 2018. A review on isolation, composition, physicochemical properties and modification of Bambara groundnut starch. *Food Hydrocolloids* 75: 62–71.

Oyeyinka, S. A., Umaru, E., Olatunde, S. J., & Joseph, J. K. 2019. Effect of short microwave heating time on physicochemical and functional properties of Bambara groundnut starch. *Food Bioscience* 28: 36–41.

Samson, A. O., Olaide, A. A., Oluwafemi, A. A., Kayitesi, E., & Njobeh, P. B. 2021. A review on the physicochemical properties of starches modified by microwave alone and in combination with other methods. *International Journal of Biological Macromolecules* 176: 87–95.

Shen, H. J., Fan, D. M., Huang, L. L., Gao, Y. S., Lian, H. Z., Zhao, J. X., & Zhang, H. 2017. Effects of microwaves on molecular arrangements in potato starch. *RSC Advances* 7: 14348–14353.

Shi, H. X., Yin, Y. Z., Wang, A. R., Fang, L. P., & Jiao, S. F. 2016. Kinetic study of the nonthermal effect of the esterification of octenyl succinic anhydride modified starch treated by microwave radiation. *Journal of Applied Polymer Science* 133: 43909.

Singh, S., Gupta, D., Jain, V., & Sharma, A. K. 2015. Microwave processing of materials and applications in manufacturing industries: A review. *Materials and Manufacturing Processes* 30 (1): 1–29.

Solaesa, Á. G., Villanueva, M., & Vela, A. J. 2022. Impact of microwave radiation on in vitro starch digestibility, structural and thermal properties of rice flour. From dry to wet treatments. *International Journal of Biological Macromolecules* 222: 1768–1777.

Souza, A. C., Benze, R. F. E. S., Ferrão, E. S., Ditchfield, C., Coelho, A. C. V., & Tadini, C. C. 2012. Cassava starch biodegradable films: Influence of glycerol and clay nanoparticles content on tensile and barrier properties and glass transition temperature. *LWT-Food Science and Technology* 46: 110–117.

Sriamornsak, P., Juttulapa, M., & Piriyaprasarth, S. 2010. Microwave-assisted modification of arrowroot starch for pharmaceutical matrix tablets. *Advanced Materials Research* 93: 358–361.

Stevenson, D. G., Biswas, A., & Inglett, G. E. 2005. Thermal and pasting properties of microwaved corn starch. *Starch-Stärke* 57: 347–353.

Sun, J., Wang, W., & Yue, Q. 2016. Review on microwave-matter interaction fundamentals and efficient microwave-associated heating strategies. *Materials* 9: 231.

Sun, Y. N., Zhang, M., & Chen, H. Z. 2020. LF NMR intelligent evaluation of rheology and printability for 3D printing of cookie dough pretreated by microwave. *LWT-Food Science and Technology* 132: 109752.

Sun, X. X., Sun, Z. Z., Saleh, A. S. M., Lu, Y. F., Zhang, X. Y., Ge, X. Z., Shen, H. S., Yu, X. Z., & Li, W. H. 2023. Effects of various microwave intensities collaborated with different cold plasma duration time on structural, physicochemical, and digestive properties of lotus root starch. *Food Chemistry* 405: 134837.

Tao, Y., Yan, B. W., Fan, D. M., Zhang, N. N., Ma, S. Y., Wang, L. Y., Wu, Y. J., Wang, M. F., Zhao, J. X., & Zhang, H. 2020. Structural changes of starch subjected to microwave heating: A review from the perspective of dielectric properties. *Trends in Food Science & Technology* 99: 593–607.

Tester, R. F., Karkalas, J., & Qi, X. 2004. Starch-composition, fine structure and architecture. *Journal of Cereal Science* 39: 151–165.

Tian, Y., Wang, Y., Herbuger, K., Petersen, B. L., Cui, Y., Blennow, A., Liu, X. X., & Zhong, Y. Y. 2023. High-pressure pasting performance and multilevel structures of short-term microwave-treated high-amylose maize starch. *Carbohydrate Polymers* 322: 121366.

Utrilla-Coello, R. G., Bello-Pérez, L. A., Vernon-Carter, E. J., Rodriguez, E., & Alvarez-Ramirez, J. 2013. Microstructure of retrograded starch: Quantification from lacunarity analysis of SEM micrographs. *Journal of Food Engineering* 116: 775–781.

Wu, F. B., Chi, B., Xu, R. Y., Liao, H. Y., Xu, X. Q., & Tan, X. Y. 2022. Changes in structures and digestibility of amylose-oleic acid complexes following microwave heat-moisture treatment. *International Journal of Biological Macromolecules* 214: 439–445.

Xia, T. Y., Gou, M., Zhang, G. Q., Li, W. H., & Jiang, H. 2018. Physical and structural properties of potato starch modified by dielectric treatment with different moisture content. *International Journal of Biological Macromolecules* 118: 1455–1462.

Xu, X. G., Chen, Y. Z., Luo, Z. G., & Lu, X. X. 2019. Different variations in structures of A- and B-type starches subjected to microwave treatment and their relationships with digestibility. *LWT-Food Science and Technology* 99: 179–187.

Xu, K., Zhang, M., & Bhandari, B. 2020. Effect of novel ultrasonic-microwave combined pretreatment on the quality of 3D printed wheat starch-papaya system. *Food Biophysics* 15: 249–260.

Yan, B. W., Shen, H. J., Fan, D. M., Tao, Y., Wu, Y. J., Wang, M. F., Zhao, J. X., & Zhang, H. 2019. Microwave treatment regulates the free volume of rice starch. *Scientific Reports* 9: 3876.

Yang, Q., Qi, L., Luo, Z. G., Kong, X. L., Xiao, Z. G., Wang, P. P., & Peng, X. C. 2017. Effect of microwave irradiation on internal molecular structure and physical properties of waxy maize starch. *Food Hydrocolloids* 69: 473–482.

Yılmaz, A., & Tugrul, N. 2023. Effect of ultrasound-microwave and microwave-ultrasound treatment on physicochemical properties of corn starch. *Ultrasonics Sonochemistry* 98: 106516.

Zeng, S. X., Chen, B. Y., Zeng, H. L., Guo, Z. B., Lu, X., Zhang, Y., & Zheng, B. D. 2016. Effect of microwave irradiation on the physicochemical and digestive properties of lotus seed starch. *Journal of Agricultural and Food Chemistry* 12: 2442–2449.

Zhang, Y. Y., Qin, Y. N., Liang, Q. R., Hu, Y. Y., & Luan, G. Z. 2023. Breaking the temperature limitation of zein-rice starch dough by microwave pre-gelatinization: Morphological, structural and rheological properties of the dough. *Food Research International* 173: 113465.

Zhang, F., Zhang, Y. Y., Thakur, K., Zhang, J. G., & Wei, Z. J. 2019. Structural and physicochemical characteristics of lycoris starch treated with different physical methods. *Food Chemistry* 275: 8–14.

Zhong, Y. J., Xiang, X. Y., Zhao, J. C., Wang, X. H., Chen, R. Y., Xu, J. G., Luo, S. J., Wu, J. Y., & Liu, C. M. 2020. Microwave pretreatment promotes the annealing modification of rice starch. *Food Chemistry* 304:125432.

Zhu, Z. Z., & Guo, W. C. 2017. Frequency, moisture content, and temperature dependent dielectric properties of potato starch related to drying with radio-frequency/microwave energy. *Scientific Reports* 7: 9311.

Zhu, J., Li, L., Zhang, S. Y., Li, X. X., & Zhang, B. J. 2016. Multi-scale structural changes of starch-based material during microwave and conventional heating. *International Journal of Biological Macromolecules* 92: 270–277.

CHAPTER 7

Ultrasound Treatment of Starch and Starchy Materials

Hebin Xu, Ankutse Peter, and Yang Tao

7.1 INTRODUCTION

The Encyclopedia Britannica defines ultrasound as vibrations of frequencies above the upper limit of the audible range for humans (greater than about 20 kHz). Ultrasound started as an ancient diagnostic medical technology. The first idea of ultrasound came from Lazzaro Spallanzani (Figure 7.1), a physicist, biologist, professor, and priest who carried out studies on bats that concluded that they could navigate using sound rather than sight. In 1877, the two brothers Pierre and Jacques Currie discovered piezoelectricity. Ultrasound transducers (probes) emit and receive sound waves by way of the piezoelectric effect. Neurologist Karl Dussik is credited with being the first to use sonography for medical diagnoses. He transmitted an ultrasound beam through the human skull in an attempt to detect brain tumors. Ultrasound application is a new technology that is currently gaining prominence. In 1994, Yoshinobu Isono, Takehisa Kumagai, and Toshiyuki Watanabe degraded waxy rice starch by ultrasound. From the 2000s up to date, this technology has become widely investigated.

Ultrasound refers to the application of high-frequency sounds to food and non-food products to manipulate, enhance, and analyze several facets of

DOI: 10.1201/9781003493594-7

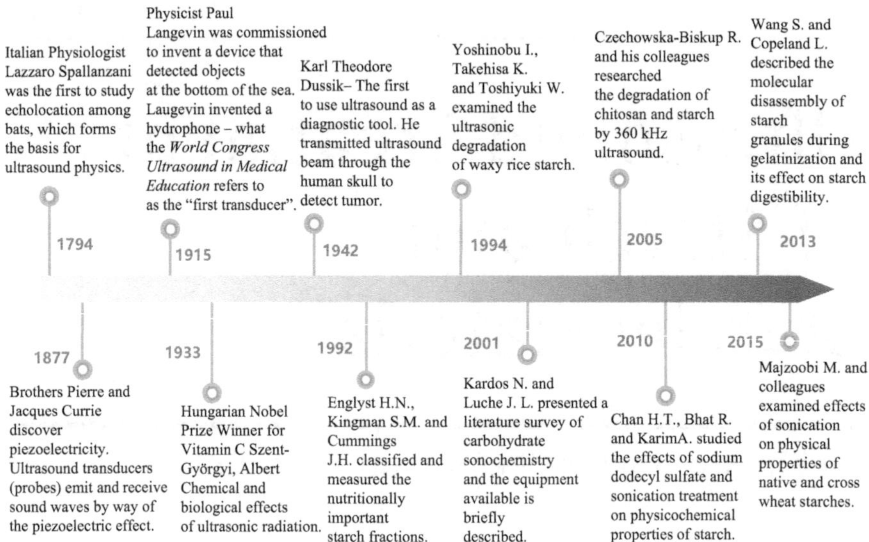

FIGURE 7.1 Development of ultrasound and its application in starch and starchy materials.

products during their production and preservation to attain a more quality control measure. Ultrasound know-how has gained acceptance in food and non-food engineering due to its non-invasive, non-destructive, and multi-purpose nature. It is a natural and physical method that has been useful in a wide area of study like nutraceutical, cosmetics, pharmaceutical, food, and chemical industries globally (Oliyaei & Moosavi-Nasab, 2021; Shalmashi, 2009; Tiwari, 2015). So far, ultrasound applications have gained prominence and have been reported in literatures to cause structural and functionality changes in starch from different origins (Han et al., 2023; Zhang et al., 2022). The application of ultrasound in the treatment of starch and starchy materials includes exposing them to auditory cavitation, localized heating, and mechanical effects produced by ultrasonic waves. This procedure encourages variations at the molecular and structural points, contributing a range of possible returns in terms of transformed rheological, thermal, and expedient characteristics. There is substantial interest in using ultrasound for diverse essential nutrition researches and commercial applications (Awad et al., 2012; Knorr et al., 2004). In this chapter, the principles of ultrasound action, structural change of ultrasound-treated starch (UTS), and the potential applications of UTS are summarized (Table 7.1).

TABLE 7.1 Reported Cases of Ultrasound Modification of the Structure of Starch and Starchy Materials

Food Type	Starch Shape	Particle Size	Temp. °C	UTS Frequency/ Amplitude	Time	Microscopy Type	Research Findings	References
Pea starch granules	Smooth & oval-like	20 μm	Under 0°C, 25°C	N/A	20 minutes	SEM, CLSM	No grooves and cracks were found at 0°C and 25°C. Slightly rough and pitted surface at temperatures above 25°C	Han et al. (2023)
Native corn starch	Smooth & oval-like	N/A	25°C	5% (400 W/cm^2)	10, 20, and 30 min	SEM	Rough surface and agglomeration of the prepared complexes. Viscosity of the CS paste decreased. The average molecular weight of the CS decreased from 380.478 to 323.989 kDa	Hao et al. (2023)
Pea starch granules (PSt) and glycerol monolaurate (GM)	Regular oval structure with smaller diameters	N/A	90°C	20% (160 W/cm^2: CU-20-10, CU-20-30), 50% (400 W/cm^2: CU-50-10, CU-50-30) and 80% (640 W/cm^2: CU-80-10, CU-80-30)	10–30 min	LE	Starch granules swelled to larger sizes. The double-helix structure of swollen starch granules opens, destroying its inherent semi-crystalline structural configuration and exposing the internal laminated structure	Zhang et al., (2022)
Cassava and corn starches	Rounded, oval, or oval-truncated	60–80 nm	26°C ± 2°C	40 kHz, 99%	10 and 20 minutes	SEM, FTIR, DSC, XRD	Groove and notch on the surface. Decreased enthalpy value, absorption peak changed	Rahaman et al., (2021)

(Continued)

TABLE 7.1 (*Continued*) Reported Cases of Ultrasound Modification of the Structure of Starch and Starchy Materials

Food Type	Starch Shape	Particle Size	Temp. °C	UTS Frequency / Amplitude	Time	Microscopy Type	Research Findings	References
Potato chips	N/A	0.4–3 μm and 7–12 μm	30°C–90°C	20 kHz (360 W, 660 W)	60 minutes	SEM	Surface erosion of starch granules and higher power made the structure of starch disorganized	Zhang et al. (2021)
Native corn starch (CS)	N/A	N/A	90°C	320 W/cm^2	30 minutes	SEM	Decrease in the Rmax and Ra values. Homogeneous structure with a smooth surface. The crystalline structure of starch was changed from the A-type to the V-type pattern	Kang et al. (2020)
Cassava tuber starch	N/A	N/A	40°C, 50°C	/	10–20 minutes	/	A slight decrease in clarity of the sonicated cassava starch paste compared to the control starch, but the differences were not very significant statistically	Krishnakumar & Sajeev (2018)
Native (N) and cross-linked (CL) wheat starches	Even and smooth	N/A	22°C	20 kHz, (100 W)	20 minutes	SEM	Cracks on the granules of both N and CL starches significantly ($P<0.05$) reduced the intrinsic viscosity of both N and CL samples indicating molecular degradation of starch	Majzoobi et al. (2015)

Scanning electron microscopy (SEM), Confocal laser scanning micrographs (CLSM), Fourier transform infrared spectroscopy (FTIR), Differential scanning calorimetry (DSC), X-ray diffraction (XRD), Light Microscope (LE), Not applicable (N/A).

7.2 PRINCIPLES OF ULTRASOUND ACTION

Ultrasound is the mechanical waves with frequencies more than 20 kHz, a limit more than human hearing capacity. It can be divided into three groups based on their frequency and power range: low-frequency high-intensity ultrasound (20–100 kHz), medium-frequency medium-intensity (100 kHz–1 MHz), and high-frequency low-intensity ultrasound (1–10 MHz) (Patist & Bates, 2008). The ultrasound used in the food industry is in the range of low and high frequency. The power used with low-intensity ultrasound is generally less than 1 $W{\cdot}cm^{-2}$, which does not affect the material structure. Inversely, low-frequency and high-intensity ultrasound with sufficient energy can cause physical damage or encourage chemical reactions like emulsion formation, crystallization control, and oxidation reactions.

Most of the ultrasonic application in starchy materials is in a liquid–solid system with water as the medium. Within the ultrasound equipment, there are mainly two components known as ultrasound transducers (Piezoelectric and Magnetostrictive). The piezoelectric component is responsible for the generation of the ultrasonic waves, while the magnetostrictive component acts as an electroacoustic transducer to aid in the generation of the ultrasonic waves. Piezoelectric devices often function at or near their accepted resonant frequency, where they can yield the maximum substantial result. This is key for applications like ultrasonic cleaning or medical imaging. However, magnetostrictive transducers act as electroacoustic transducers and are responsible for the creation of ultrasonic waves on the principle of magnetostriction which is described as the successive variation in length per unit length (Bhargava et al., 2021). In most reported studies, an ultrasonic probe system is used for the treatment of starch as shown in Figure 7.2. The probe receives energy from the

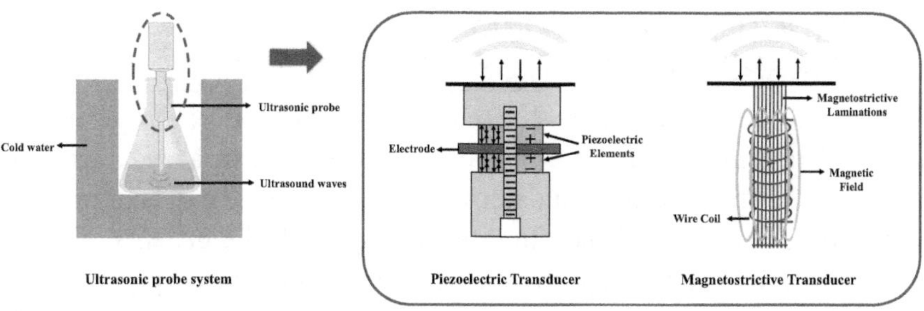

FIGURE 7.2 Ultrasound system usually applied in starch and starchy materials.

transducer and transmits the energy to the sample. The amount of starch that can be effectively processed is determined by the diameter of the probe tip. During sonication, the temperature of the system will increase, which can be controlled by adding ice to the bath system.

The cavitation effect and radical attack are the two primary mechanisms governing the changes in the characteristics of starch during the ultrasonic process. The sinusoidal waves passing through the medium cause a continuous wave-type motion and the changing pressure from the motion results in mixing within the liquid. When the local pressure in the medium is lower than the vapor pressure of the liquid, tiny bubbles can be created. The bubbles keep growing during the expansion cycle and shrink during the compression cycle. Over numerous cycles, the bubbles keep growing. When the oscillation of the bubble wall equals that of the applied frequency of the sound waves, the bubbles explode in one compression cycle. The collapse of the bubbles caused by this process contributes to the cavitation effect (Patist & Bates, 2008). The generation of gas bubbles starts first, followed by producing intense micro-jets with velocities higher than 100 m/s which causes the collide of the particles. The rapid collapse of each microbubble generates energy to increase high pressure (1000 atm) and temperature (4000 K), modifying the physical and chemical conditions of the system. Moreover, the shear forces arise due to rapidly collapsing bubbles and disrupt the polymer chains of starch. In conclusion, the micro-jets and shear forces induced by cavitation are responsible for the damage to the starch granules (Zuo et al., 2012). When it comes to radical attack, free radicals such as hydroxide (•OH) and (•H) hydrogen radicals may be generated from the bubbles collapsing (Czechowska-Biskup et al., 2005). A radical attack occurs when water is partially decomposed into hydroxide and hydrogen radicals, causing starch polymer breakdown and further contributing to the change of physicochemical properties of starch.

7.3 STRUCTURAL CHANGE OF ULTRASOUND-TREATED STARCH (UTS)

Ultrasound treatment alters the physical and chemical properties of starch, rendering it extra multipurpose and suitable in a wide array of products. Relatively, ultrasound know-how is considered to be natural and innovative, non-thermal handling technology with the benefits of rapid accomplishment time, high proficiency, zero pollution, easy procedure, and little energy consumption, which is extensively used to modify starch (Hao et al., 2023). Besides, these changes may present successions of variations and exchanges, with a momentous influence on the value of the final

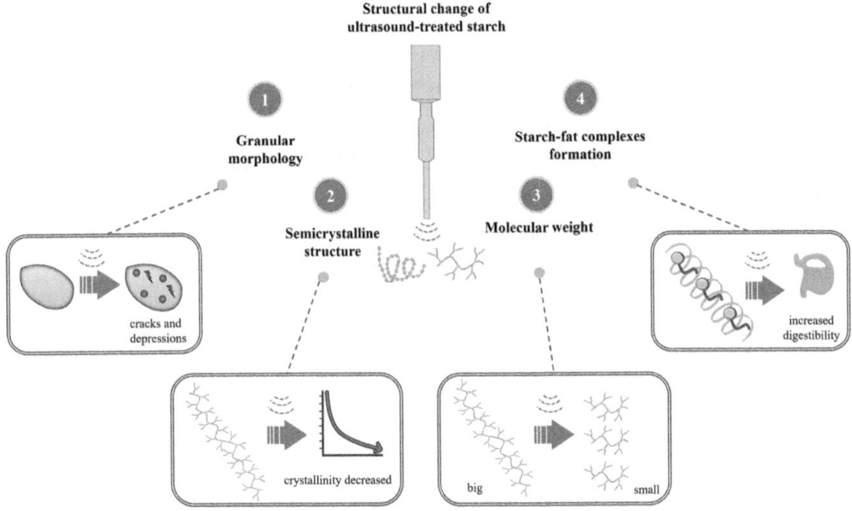

FIGURE 7.3 Structural change of ultrasound-treated starch.

product like texture, mouthfeel, flavor, and nutritional traits (Parada & Santos, 2016; Wang & Copeland, 2013). Structural modifications of starch can be achieved through ultrasound treatment and help to improve the physicochemical properties of starch (He et al., 2023). The structural change of UTS is shown in Figure 7.3.

7.3.1 Granular Morphology

The microscopic effects of ultrasound on the structure of potato starch were studied and resulted in increased vaporous cavitation, and many deep conical holes on the starch granules (Gallant et al., 1972). In some experiments, the application of ultrasonic has resulted in cracks and depressions on the surfaces of corn, wheat, potato, and rice starch granules (Sujka & Jamroz, 2013). Potato starches were reported to display a characteristic spherical and ellipsoid shape, corroborating the shape of potato starch granules reported by (Dhital et al., 2010). Despite the natural ellipsoidal shape, ultrasonication was reported to cause bulges on the surfaces of the potato starch granule (Zhang et al., 2005). The bulginess of the potato starch granules may be attributed to the biochemical processes of building starch granules by converting photosynthesis into glucose (Chen et al., 2019). In peas, sweet potato, and potato starch, at low and high temperatures (i.e., 25°C and 50°C), ultrasonication increased the number of pores and cracks on the starch granule surface (Chan et al., 2021). Ultrasound-treated corn starch granules were observed at 40 kHz by scanning electron microscopy (SEM), Feourier transform infrared spectroscopy (FTIR), differential scanning calorimetry (DSC), and X-ray diffraction (XRD). Grooves and notches appeared on the surface of the starch granules (Rahaman et al., 2021). Sujka (2017) reported a change in the morphological structure of native rice starch by ultrasound treatment. The treated rice in ethanol displayed a weighty increase in the average diameter of mesopores and exhibited novel pores with diameter ranges from 1.7 to 300 nm (Rahaman et al., 2021).

7.3.2 Semi-Crystalline Structure

Insights into the molecular structure of pea starch revealed by the use of Fourier transform infrared spectroscopy and small-angle X-ray scattering analyses discovered that UT lessens the short-range group and improves the width of semi-crystalline and amorphous lamellae by encouraging starch chain depolymerization. The pea starch treated by ultrasound at 45°C had the larger part of B2 chains associated with the other ultrasound-treated samples since the higher ultrasonic temperature altered the disruption sites of starch chains (Han et al., 2023). The impact of multicycle ultrasound-assisted ice recrystallization (US+IR) joint with amyloglucosidase (AMG) or maltogenic α-amylase (MA) catalyzed hydrolysis on structure was investigated. Scanning electron microscopy (SEM) displayed that the US+IR produced shallow indents and trenches on the exterior of granules while the combination US+IR and enzyme hydrolysis created additional pores on starch granules. The US+IR treatment meaningly decreased the relative crystallinity, amylose portion, and bulge size (Keeratiburana et al., 2020). Also, the crystalline structure of native corn and cassava starches was investigated and the XRD outcomes discovered a minor decrease in the crystallinity degree (CD) of sonicated corn (25.3, 25.1) and cassava starch (21.0, 21.4) as compared to native corn (25.6%) and cassava starch (22.2%) (Rahaman et al., 2021). Yang et al. (2019) reported a declining trend of relative crystallinity of rice starch when treated by 150, 300, 450, and 600 W and the results 31.24%, 33.65%, 27.09%, and 26.24%, respectively, lower than that of the native counterpart (34.60%). Occasionally, Sonication treatment for lengthier time or at higher power reduced α-helix and β-turn while increasing β-sheet and random coil ($p < 0.05$) (Liu et al., 2021). Studies by Liu et al. (2021) have revealed that ultrasonication affects the crystallinity of maize starch granules. The treatment can lead to either an upsurge or a reduction in crystallinity depending on the processing conditions. Deviations in crystallinity are strictly correlated with the functional properties of starch, mostly in terms of its digestibility and vulnerability to enzymatic degradation (Liu et al., 2021).

Novel lotus root starch (LRS)-myristic acid (MA) complexes were modified by an ultrasound-assisted hydrothermal method (UHM) to study its nourishing involvement in type II diabetes mellitus (T2DM). In that experiment, ultrasonic treatment encouraged the development of the V-type crystal structure complex and enhanced the intermolecular contact force, the arrangement of the short-range starch molecules, and crystallinity (Tang et al., 2022). There is a link between starch lipids and crystalline structure formation. For instance, the complexation of starch and lipids controlled the conversion of the crystal structure of lotus root starch from A-type to V-type (Tang et al., 2022). In the cooling and recrystallization, pure amylose forms B-type crystals; nevertheless, due to the access of guest molecules, amylose guest molecule complexation would strive with amylose recrystallization to form V-type crystals (Chen et al., 2019). Food preservation is significant in avoiding the deterioration of food substance. Ultrasonication is reported to be used to crystallize food products such as triglyceride oils (Kiani & Sun, 2018), Milk (Martini et al., 2008), and ice cream (Mortazavi & Yazdi, 2008).

7.3.3 Molecular Weight

Ultrasound can be used to hydrolyze starch into smaller fragments, such as maltodextrins or glucose. This is significant in the manufacture of syrups, sweeteners, and other starch-derived products. The ultrasonic energy may have a slight effect on minor molecules but has the capacity to disrupt the chains of large polymers like starch (Arzeni et al., 2012; Chan et al., 2010). The assessment of pasting properties of novel native starch derived from defatted and depigmented annatto seeds showed a product with a peak viscosity larger than corn starch, which is a significant feature for texture classification. Additionally, the pasting temperature was 7.9°C less, permitting its application in the production of functional diets (Zabot et al., 2019). Sometimes, the original starch can be dextrinized with entwined macromolecular chains which present a very high viscosity thereby resisting easy flow. Application of ultrasound techniques is required to break the chains of macro-molecules of the starch hence an alteration in the crystalline structure thus lowering the viscosity (Sharma et al., 2015; Warren et al., 2016). Occasionally the transparency of starch paste is mirrored by its transmittance. During the sonication time of 30 min, the light transmission improved to 38.55% because of the disintegration of the amorphous and minor crystalline regions of the starch granules at this point, which also permitted the starch chains to be more entirely distributed in the paste, dropping the reproduction of light and thus increasing the light transmission (Hao et al., 2023). The thermal and pasting characteristics of rice starch were also determined by DSC and RVA. The outcomes revealed that ultrasound treatments cause the peak and breakdown viscosity to increase, although pasting temperature, gelatinization enthalpy, and the peak time decreased (Yang et al., 2019).

7.3.4 Starch-fat Complexes Formation

The complexes of starch-lipid by ultrasound treatment were studied with the interactions that occurred using molecular docking. The scanning electron microscope (SEM) outcomes exposed a rough surface and agglomeration of the prepared complexes, and the complexes of corn starch (CS)-lauric acid (LA) improved by 14.03% when compared to the non-ultrasound group (Hao et al., 2023). As an advantage, starch-lipid complexes lessen the bulge power and the ability of the starch to be soluble, retrogradable, gelatinizable, and its rate of enzymic digestion (Copeland et al., 2009; Putseys et al., 2010; Wang & Copeland, 2013). While the molecular weight of CS decreased from 380.478 to 323.989 kDa, an extra well-organized helical structure and a more solid V-shaped crystal structure over hydrophobic exchanges and hydrogen bonds, provided a foundation for the association between structure and digestibility of lipid-containing starchy foods (Hao et al., 2023). Tang et al. (2022) investigated the dietary supplements of 5% and 15% ultrasound novel lotus root starch (LRS)-myristic acid (MA) (U-LRS-MA). U-LRS-MA significantly lessens the organ index, the body mass, and the fasting blood glucose of type II diabetes mellitus (T2DM) mice, thereby successfully regulating its blood lipid level, improving its liver injury, and improving the stages of colonic short-chain fatty acids. Conclusively, the use of 5% U-LRS-MA was more effective in T2DM than 15% U-LRS-MA. Ultrasound could be efficiently applied to formulate lipid-starch complexes, specifically type 5 resistant starch, which was evidenced for the first time to have an outstanding intervention outcome on T2DM.

7.4 UTS APPLICATIONS

Ultrasound sonification technology has been widely accepted in several fields like the nutraceutical, cosmetics, pharmaceutical, food, and chemical industries globally (Oliyaei & Moosavi-Nasab, 2021; Shalmashi, 2009; Tiwari, 2015), due to its natural and physical applications (Dzah, Duan, Zhang, Wen, et al., 2020). UTS, or sonicated starch, has gotten attention

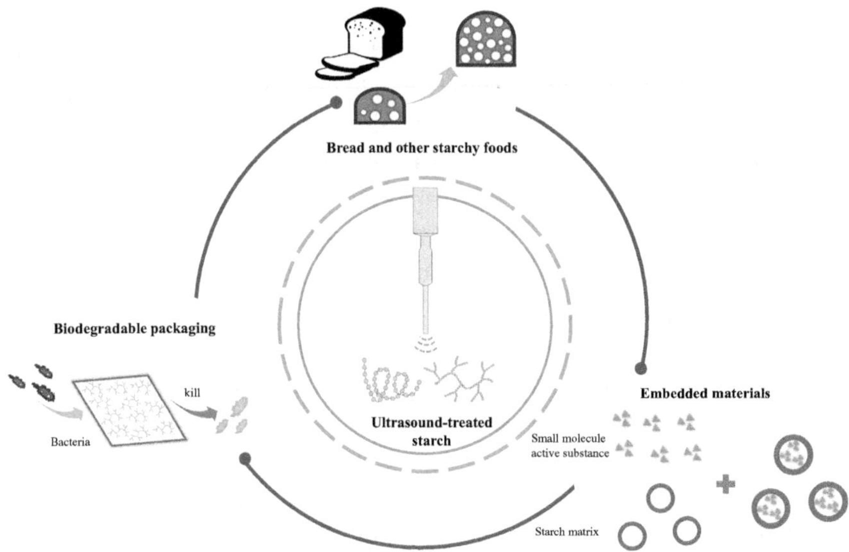

FIGURE 7.4 UTS applications in food and other industries.

in numerous trades owing to its exceptional qualities and possible applications. Many studies have revealed that the UTS has enormous potential for various food products and starch-based packaging applications, which is illustrated in Figure 7.4.

7.4.1 Bread and Other Starchy Foods

After modification, the reduced moisture regains viscosity and can be used to improve the quality of other starch foods such as bread. In Veal's study, ultrasound-treated rice flour was used to replace 30% native flour in gluten-free bread. They employed doughs' pasting, thermal, and rheological properties to characterize the production performance of the bread. Ultrasound showed to significantly improve the fermentative and viscoelastic behavior of rice flour dough and its bread-making performance (Vela et al., 2023). Qin et al. (2022) applied ultrasound to modified rice flour. The ultrasound treatment significantly changed the physicochemical properties of rice flour and further changed the rice bread quality. The viscoelasticity of the batter containing 15% UTS was significantly improved compared to the semidry-milled rice flour batter, leading to a higher bread volume and softer crumb texture (Qin et al., 2022). Thus, ultrasound modification is an alternative to improve the morphology and physical properties of starchy materials and further improve bread quality.

7.4.2 Biodegradable Packaging

The development of alternative materials for biodegradable packaging has been recently the subject of much research to prevent environmental pollution caused by packaging waste. Edible films and coatings produced from various biopolymers have gained great interest since they provide a good barrier against oxidative and physical stress in biodegradable packaging materials (Mohamed et al., 2020). Starch, as an inexpensive and biodegradable edible material that has a good film-forming ability, is considered one of the most promising alternatives to packaging. However, edible films made of starch have some drawbacks, such as high water penetration and poor mechanical characteristics. The films produced using ultrasonicated starch showed a more cohesive structure, improving the mechanical and barrier properties (Chen et al., 2021). Rahmasari and Yemiş (2022) developed a novel edible ginger starch-based film incorporated with coconut shell liquid smoke (CSLS) by ultrasonication. They declared that UTS films presented better mechanical, barrier, thermal, and antibacterial properties compared to control films. Ultrasound-treated CSLS-ginger starch films appeared to be promising for active antimicrobial packaging applications of meat products (Rahmasari & Yemiş, 2022).

7.4.3 Embedded Materials

UTS also received great interest in biomaterials as a carrier of bioactive substances, thus exhibiting various applications in chemical, food, pharmaceutical, cosmetic industries, and so on. Cinnamaldehyde is a food antimicrobial agent that can be used to prolong the shelf life (Gutiérrez et al., 2009). Nevertheless, the irritating smell of cinnamaldehyde affects the intrinsic flavor of foods, and large quantities of active ingredients are also neutralized by free cinnamaldehyde in foods. Some scholars employed starch materials to encapsulate cinnamaldehyde to improve its stability (Jiang et al., 2010; Ponce Cevallos et al., 2010). High-amylose starch with ultrasound treatment is one of the most advantageous techniques for the encapsulation of cinnamaldehyde. The ultrasound treatment could reduce the molecular weight of starch and increase the production yield of the inclusion complex. Cinnamaldehyde was successfully encapsulated by high-amylose corn starch to form starch–cinnamaldehyde inclusion complex. The encapsulation increased the stability of cinnamaldehyde and generated a good releasing behavior in food preservation. Tangeretin has multiple biological activities, such as antioxidant, anti-cancer, and anti-inflammatory properties, as well as antidiabetic effects (Lee et al., 2016; Wunpathe et al., 2020). However, the weak stability and the poor aqueous solubility of tangeretin severely restrict its application in industries. Hu et al. applied a green technique comprised of a combination of ultrasound and recrystallization. Debranched starch and β-cyclodextrin underwent ultrasonic irradiation and loaded tangeretin. After 20 days of storage, the maximum retention index of starch-coated tangeretin was 99.21% ± 0.17%. The prepared nanocarriers were demonstrated to be non-toxic to cells, showing promising potential for applications in the food, pharmaceutical, and cosmetic industries (Hu et al., 2020).

Generally, ultrasound treatment of starch holds great promise for addressing industry challenges, improving product quality, thus contributing to sustainable and innovative practices across diverse applications. In the future, more and more research will focus on the deeper mechanisms of UTS, such as exploring the difference between modified starches from different sources. Consumers will experience the convenience of UTS in various fields, such as food, pharmaceutical, and environmental protection. Ongoing research and technological improvements are likely to unlock innovative potentials and increase the range of commercial applications for ultrasound-modified starch and starchy materials.

In conclusion, the ultrasound treatment of starch and starchy materials signifies a dynamic and evolving sector with substantial suggestions for diverse industries. The current research indicates that ultrasound can effectively modify the structural and functional properties of starch, offering opportunities to enhance productivity, customize functionalities, and create inventive products. As technology continues to advance, the future prospects for ultrasound-assisted starch modification are promising, encircling cleaner and more justifiable processing methods, the advance of novel food and pharmaceutical applications, and integration with evolving technologies. The constant investigation of ultrasound's potential in starch modification highlights its role as a multipurpose tool for tailoring materials to meet the evolving demands of various sectors.

REFERENCE, BIBLIOGRAPHY OR WORKS CITED

Arzeni, C., Martínez, K., Zema, P., Arias, A., Pérez, O. E., & Pilosof, A. M. R. 2012. Comparative study of high intensity ultrasound effects on food proteins functionality. *Journal of Food Engineering* 108 (3): 463–472.

Awad, T. S., Moharram, H. A., Shaltout, O. E., Asker, D., & Youssef, M. M. 2012. Applications of ultrasound in analysis, processing and quality control of food: A review. *Food Research International* 48 (2): 410–427.

Bhargava, N., Mor, R. S., Kumar, K., & Sharanagat, V. S. 2021. Advances in application of ultrasound in food processing: A review. *Ultrasonics Sonochemistry* 70: 105293.

Chan, H. T., Bhat, R., & Karim, A. A. 2010. Effects of sodium dodecyl sulphate and sonication treatment on physicochemical properties of starch. *Food Chemistry* 120 (3): 703–709.

Chan, C. H., Wu, R. G., & Shao, Y. Y. 2021. The effects of ultrasonic treatment on physicochemical properties and in vitro digestibility of semigelatinized high amylose maize starch. *Food Hydrocolloids* 119: 106831.

Chen, W., Chao, C., Yu, J., Copeland, L., Wang, S., & Wang, S. 2021. Effect of protein-fatty acid interactions on the formation of starch-lipid-protein complexes. *Food Chemistry* 364: 130390.

Chen, L., Ma, R., Zhang, Z., Huang, M., Cai, C., Zhang, R., McClements, D. J., Tian, Y., & Jin, Z. 2019. Comprehensive investigation and comparison of surface microstructure of fractionated potato starches. *Food Hydrocolloids* 89: 11–19.

Copeland, L., Blazek, J., Salman, H., & Tang, M. C. 2009. Form and functionality of starch. *Food Hydrocolloids* 23 (6): 1527–1534.

Czechowska-Biskup, R., Rokita, B., Lotfy, S., Ulanski, P., & Rosiak, J. M. 2005. Degradation of chitosan and starch by 360-kHz ultrasound. *Carbohydrate Polymers* 60 (2): 175–184.

Dhital, S., Shrestha, A. K., & Gidley, M. J. 2010. Relationship between granule size and in vitro digestibility of maize and potato starches. *Carbohydrate Polymers* 82 (2): 480–488.

Gallant, D., Degrois, M., Sterling, C., & Guilbot, A. 1972. Microscopic effects of ultrasound on the structure of potato starch preliminary study. *Starch-Stärke* 24 (4): 116–123.

Gutiérrez, L., Escudero, A., Batlle, R., & Nerín, C. 2009. Effect of mixed antimicrobial agents and flavors in active packaging films. *Journal of Agricultural and Food Chemistry* 57 (18): 8564–8571.

Han, L., Huang, J., Yu, Y., Thakur, K., Wei, Z., Xiao, L., & Yang, X. 2023. The alterations in granule, shell, blocklets, and molecular structure of pea starch induced by ultrasound. *International Journal of Biological Macromolecules* 240: 124319.

Hao, Z., Xu, H., Yu, Y., Han, S., Gu, Z., Wang, Y., Li, C., Zhang, Q., Deng, C., Xiao, Y., Liu, Y., Liu, K., Zheng, M., Zhou, Y., & Yu, Z. 2023. Preparation of the starch-lipid complexes by ultrasound treatment: Exploring the interactions using molecular docking. *International Journal of Biological Macromolecules* 237: 124187.

He, R., Li, S., Zhao, G., Zhai, L., Qin, P., & Yang, L. 2023. Starch modification with molecular transformation, physicochemical characteristics, and industrial usability: A state-of-the-art review. *Polymers* 15 (13): 2935.

Hu, Y., Qin, Y., Qiu, C., Xu, X., Jin, Z., & Wang, J. 2020. Ultrasound-assisted self-assembly of β-cyclodextrin/debranched starch nanoparticles as promising carriers of tangeretin. *Food Hydrocolloids* 108: 106021.

Jiang, S., Li, J. N., & Jiang, Z. T. 2010. Inclusion reactions of β-cyclodextrin and its derivatives with cinnamaldehyde in *Cinnamomum loureirii* essential oil. *European Food Research and Technology* 230 (4): 543–550.

Kang, X., Liu, P., Gao, W., Wu, Z., Yu, B., Wang, R., Cui, B., Qiu, L., & Sun, C. 2020. Preparation of starch-lipid complex by ultrasonication and its film forming capacity. *Food Hydrocolloids* 99: 105340.

Keeratiburana, T., Hansen, A. R., Soontaranon, S., Tongta, S., & Blennow, A. 2020. Porous rice starch produced by combined ultrasound-assisted ice recrystallization and enzymatic hydrolysis. *International Journal of Biological Macromolecules* 145: 100–107.

Kiani, H., & Sun, D. W. 2018. Numerical simulation of heat transfer and phase change during freezing of potatoes with different shapes at the presence or absence of ultrasound irradiation. *Heat and Mass Transfer* 54 (3): 885–894.

Knorr, D., Zenker, M., Heinz, V., & Lee, D. U. 2004. Applications and potential of ultrasonics in food processing. *Trends in Food Science and Technology* 15 (5): 261–266.

Krishnakumar, T., & Sajeev, M. 2018. Effect of ultrasound treatment on physicochemical and functional properties of cassava starch. *International Journal of Current Microbiology and Applied Sciences* 7 (10): 3122–3135.

Lee, Y. Y., Lee, E. J., Park, J. S., Jang, S. E., Kim, D. H., & Kim, H. S. (2016). Anti-inflammatory and antioxidant mechanism of tangeretin in activated microglia. *Journal of Neuroimmune Pharmacology* 11 (2): 294–305.

Liu, P., Gao, W., Zhang, X., Wang, B., Zou, F., Yu, B., Lu, L., Fang, Y., Wu, Z., Yuan, C., & Cui, B. 2021. Effects of ultrasonication on the properties of maize starch/ stearic acid/ sodium carboxymethyl cellulose composite film. *Ultrasonics Sonochemistry* 72: 105447.

Majzoobi, M., Seifzadeh, N., Farahnaky, A., & Mesbahi, G. 2015. Effects of sonication on physical properties of native and cross-linked wheat starches. *Journal of Texture Studies* 46 (2): 105–112.

Martini, S., Suzuki, A. H., & Hartel, R. W. 2008. Effect of high intensity ultrasound on crystallization behavior of anhydrous milk fat. *Journal of the American Oil Chemists' Society* 85 (7): 621–628.

Mohamed, S. A. A., El-Sakhawy, M., & El-Sakhawy, M. A. M. 2020. Polysaccharides, protein and lipid-based natural edible films in food packaging: A review. *Carbohydrate Polymers* 238: 116178.

Mortazavi, S. A. R., & Yazdi, F. T. 2008. Study of ice cream freezing process after treatment with ultrasound. *World Applied Sciences Journal* 4 (2): 188–190

Oliyaei, N., & Moosavi-Nasab, M. 2021. Ultrasound-assisted extraction of fucoxanthin from *Sargassum angustifolium* and *Cystoseira indica* brown algae. *Journal of Food Processing and Preservation* 45 (11): e15929.

Parada, J., & Santos, J. L. 2016. Interactions between starch, lipids, and proteins in foods: Microstructure control for glycemic response modulation. *Critical Reviews in Food Scicience and Nutrition* 56 (14): 2362–2369.

Patist, A., & Bates, D. 2008. Ultrasonic innovations in the food industry: From the laboratory to commercial production. *Innovative Food Science and Emerging Technologies* 9 (2): 147–154.

Ponce Cevallos, P. A., Buera, M. P., & Elizalde, B. E. 2010. Encapsulation of cinnamon and thyme essential oils components (cinnamaldehyde and thymol) in β-cyclodextrin: Effect of interactions with water on complex stability. *Journal of Food Engineering* 99 (1): 70–75.

Putseys, J. A., Lamberts, L., & Delcour, J. A. 2010. Amylose-inclusion complexes: Formation, identity and physico-chemical properties. *Journal of Cereal Science* 51 (3): 238–247.

Qin, W., Xi, H., Wang, A., Gong, X., Chen, Z., He, Y., Wang, L., Liu, L., Wang, F., & Tong, L. 2022. Ultrasound treatment enhanced semidry-milled rice flour properties and gluten-free rice bread quality. *Molecules* 27 (17): 5403.

Rahaman, A., Kumari, A., Zeng, X.-A., Adil Farooq, M., Siddique, R., Khalifa, I., Siddeeg, A., Ali, M., & Faisal Manzoor, M. 2021. Ultrasound based modification and structural-functional analysis of corn and cassava starch. *Ultrasonics Sonochemistry* 80: 105795.

Rahmasari, Y., & Yemiş, G. P. 2022. Characterization of ginger starch-based edible films incorporated with coconut shell liquid smoke by ultrasound treatment and application for ground beef. *Meat Science* 188: 108799.

Shalmashi, A. 2009. Ultrasound-assisited extraction of oil from the seed. *Journal of Food Lipids* 16 (4): 465–474.

Sharma, M., Yadav, D. N., Singh, A. K., & Tomar, S. K. 2015. Rheological and functional properties of heat moisture treated pearl millet starch. *Journal of Food Science and Technology* 52 (10): 6502–6510.

Sujka, M. 2017. Ultrasonic modification of starch-impact on granules porosity. *Ultrasonics Sonochemistry* 37: 424–429.

Sujka, M., & Jamroz, J. 2013. Ultrasound-treated starch: SEM and TEM imaging, and functional behaviour. *Food Hydrocolloids* 31 (2): 413–419.

Tang, J., Liang, Q., Ren, X., Raza, H., & Ma, H. 2022. Insights into ultrasound-induced starch-lipid complexes to understand physicochemical and nutritional interventions. *International Journal of Biological Macromolecules* 222: 950–960.

Tiwari, B. K. 2015. Ultrasound: A clean, green extraction technology. *TrAC Trends in Analytical Chemistry* 71: 100–109.

Vela, A. J., Villanueva, M., & Ronda, F. 2023. Physical modification caused by acoustic cavitation improves rice flour bread-making performance. *LWT* 183: 114950.

Wang, S., & Copeland, L. 2013. Molecular disassembly of starch granules during gelatinization and its effect on starch digestibility: A review. *Food and Function* 4 (11): 1564–1580.

Warren, F. J., Gidley, M. J., & Flanagan, B. M. 2016. Infrared spectroscopy as a tool to characterise starch ordered structure-a joint FTIR-ATR, NMR, XRD and DSC study. *Carbohydrate Polymers* 139: 35–42.

Wunpathe, C., Maneesai, P., Rattanakanokchai, S., Bunbupha, S., Kukongviriyapan, U., Tong-un, T., & Pakdeechote, P. 2020. Tangeretin mitigates l-NAME-induced ventricular dysfunction and remodeling through the AT1R/pERK1/2/pJNK signaling pathway in rats. *Food and Function* 11(2): 1322–1333.

Yang, W., Kong, X., Zheng, Y., Sun, W., Chen, S., Liu, D., Zhang, H., Fang, H., Tian, J., & Ye, X. 2019. Controlled ultrasound treatments modify the morphology and physical properties of rice starch rather than the fine structure. *Ultrasonics Sonochemistry* 59: 104709.

Zabot, G. L., Silva, E. K., Emerick, L. B., Felisberto, M. H. F., Clerici, M. T. P. S., & Meireles, M. A. A. 2019. Physicochemical, morphological, thermal and pasting properties of a novel native starch obtained from annatto seeds. *Food Hydrocolloids* 89: 321–329.

Zhang, J., Yu, P., Fan, L., & Sun, Y. 2021. Effects of ultrasound treatment on the starch properties and oil absorption of potato chips. *Ultrasonics Sonochemistry* 70: 105347.

Zhang, X., Mi, T., Gao, W., Wu, Z., Yuan, C., Cui, B., Dai, Y., & Liu, P. 2022. Ultrasonication effects on physicochemical properties of starch-lipid complex. *Food Chemistry* 388: 133054.

Zhang, Z., Niu, Y., Eckhoff, S. R., & Feng, H. 2005. Sonication enhanced cornstarch separation. *Starch-Stärke* 57 (6): 240–245.

Zuo, Y. Y. J., Hébraud, P., Hemar, Y., & Ashokkumar, M. 2012. Quantification of high-power ultrasound induced damage on potato starch granules using light microscopy. *Ultrasonics Sonochemistry* 19 (3): 421–426.

CHAPTER 8

Cold Plasma Treatment of Starch and Starchy Materials

Qingqing Zhu, Enbo Xu, Huifang Cao, Xiuqin Chen, and Tian Ding

8.1 INTRODUCTION

Cold plasma (CP) as non-thermal plasma in an emerging non-thermal discharge processing is a partially ionized gas produced by applying sufficient energy to a neutral gas at low or atmosphere pressure (Moreau et al., 2008). When the gas discharge pressure is at or near atmospheric pressure, the collisions between gas molecules and electrons are less frequent, and the temperature of the plasma formed is generally around room temperature. The ionization of CP is about a maximum temperature of a few hundred kelvin (Šimončicová et al., 2019). It has a new history of use and application in different fields (Figure 8.1), such as starch modification (Zhang et al., 2022), shelf-life extension (Moradi et al., 2023), cooking optimization (Thirumdas et al., 2016), seed germination (Dhakal et al., 2023), and microbial inactivation (Li et al., 2023).

Starch, as a natural macromolecule and renewable natural polysaccharide biopolymer, is usually modified by plasma to alter its structure for functionalities (Okyere et al., 2022). Due to the spatial alignment of amylose and amylopectin, the multiscale structure of starch is complicated, and

 DOI: 10.1201/9781003493594-8

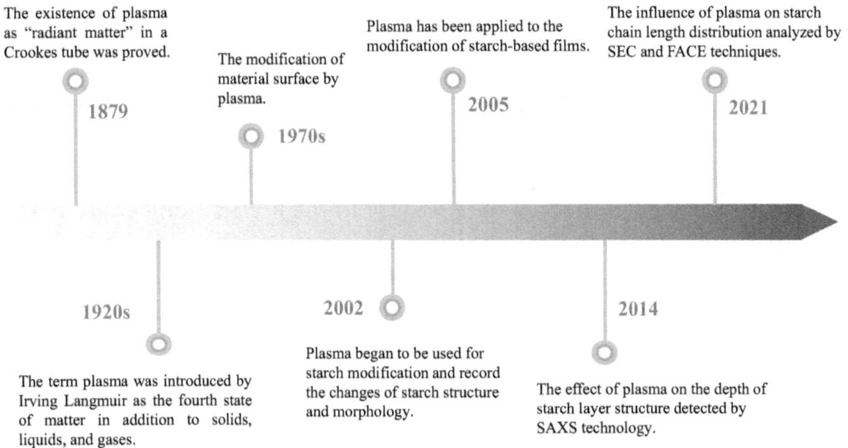

FIGURE 8.1 Timeline of important events held in the field of the starch process via cold plasma.

starch modification has received a lot of attention because of its potential processability, high biosafety, and high degradability (Kumari & Sit, 2023). It is widely used as food or a food-scale additive, as well as in green packaging and other industries. When starch granules are subjected to CP treatment, the distribution of the starch chain length is altered. This also has an impact on the starch multi-scale structures, such as crystal zones and lamellar structures, as well as on the morphology of the starch granules, which has an impact on applications based on starch for both food and non-food products. CP is non-toxic (green by-products), and always exhibits protective effects for heat-sensitive compounds (whether in the endogenous or exogenous state to starch matrix) when compared to conventional methods of starch modification.

At present, gelatinization, phase transitions, viscosity, solubility, and swelling power are the primary physical and functional qualities that are changed when starch is modified using CP. The application of modified starch is challenging to distinguish from the controllable changes in starch-based products, and it is unclear how the active matter generated by CP supports broad modifications in starch structure. Thus, this chapter shows the investigation of CP-modified starch recently and offers a new perspective for better understanding the modified structure and property involved in starch processing in various CP modes.

8.2 PRINCIPLE OF COLD PLASMA ACTION

Plasma, as the fourth matter state that exists, could be produced artificially by applying energy to gas, which dissociate electrons from gas molecules, forming ions and other excited species (Sreedevi & Suresh, 2023). Ionization is caused by these ions, electrons, and gas molecules colliding at a frequency determined by the pressure in the area (Zhong et al., 2019). Atomic oxygen (O), superoxide anions (O_2^-), singlet oxygen ($^1O_2$), ozone (O_3), hydronium ion (H_3O), hydroxyl (OH), hydrogen peroxide (H_2O_2), hydroperoxy radical (HO_2) are the common reactive oxygen species produced during plasma discharge (Kavitha et al., 2021). Nitric oxide (NO), nitrite (NO_2), nitrate (NO_3), nitrous acid (HNO_2), nitric acid (HNO_3), and peroxynitrous acid (ONOOH) are the reactive nitrogen species (RNS) that are produced (Jha et al., 2017).

As is shown in Figure 8.2, the most common CP generation methods used in starch processing are dielectric barrier discharge plasma (DBDP), glow discharge plasma (GDP), radiofrequency plasma (RFP), atmospheric pressure plasma jet (APPJ), and corona electrical discharge plasma (CEDP). The differentiated effects of DBDP, GDP, RFP, JP, and CEDP patterns on starch-based objectives are summarized in Section 8.2. Reactive radical formation can be controlled and managed both qualitatively and quantitatively by manipulating the gas flow rate, pressure exerted, choice of plasma-forming gas, discharge voltage, and treatment/exposure duration (Meiyazhagan et al., 2020).

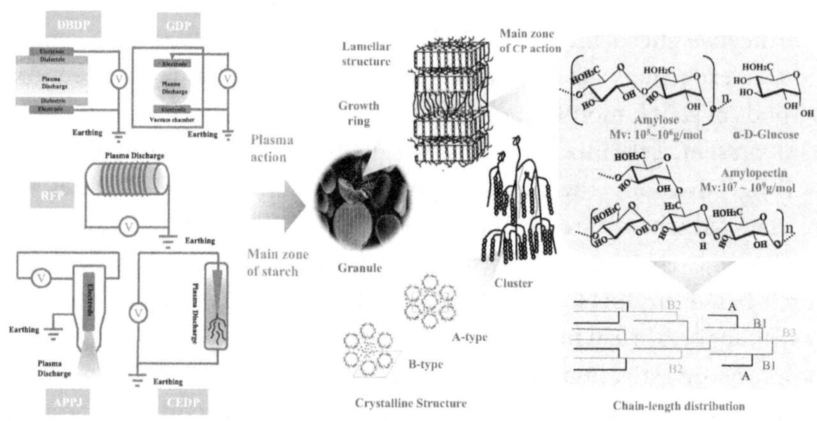

FIGURE 8.2 Five plasmas (DBDP, GDP, RFP, APPJ, and CEDP) commonly used for starch modification and starch action sites.

8.2.1 Dielectric Barrier Discharge Plasma (DBDP)

DBDP, as the most commonly used plasma generation mode for starch and starch-based film modification, is typically generated via inserting one or more dielectrics between the high voltage and the ground electrode (Li et al., 2020). The simplicity of the uniform discharge and its stable and reproducible plasma conditions as well as the flexibility of the dielectric materials (glass, quartz, ceramic, polymer layers, etc.) facilitate that plasma is widely used in scientific researches and industrial applications (Xiao et al., 2014).

8.2.2 Glow Discharge Plasma (GDP)

GDP is sustained by direct current around the pointed metallic anode or a small hole on the diaphragm when the air pressure is <100,00 Pa and the temperature is <7,00 K (Liu et al., 2022; Wang et al., 2012). GDP has the advantages of good discharge uniformity, moderate electron density, and high energy density. Moreover, it is more suitable for the surface modification of materials than other types of discharge. The current intensity of GDP is only a few milliamps, and the discharge gas temperature is low, which could reduce the sputtering rate and/or minimize thermal damage to the starch surface (Wagatsuma, 2020).

8.2.3 Radiofrequency Plasma (RFP)

RFP is formed by an oscillating magnetic field, which can ionize the gas and produces plasma (Chizoba Ekezie et al., 2017). The frequency of alternating current (AC) voltage used is typically in the radiofrequency (RF) range (1 kHZ–10^3 MHz; the most common value in starch modification is 13.56 MHz). There are three kinds of RFP discharge according to antenna coupling and the most common one is using electromagnetic wave components to provide energy to the plasma, which has the advantages of high energy density, high efficiency, uniformity, and stability (Saremnezhad et al., 2021).

8.2.4 Atmospheric Pressure Plasma Jet (APPJ)

APPJ is a discharge state when electricity is supplied to the incoming gas and active particles are created to the effluent by the air stream (Sigeneger et al., 2017). APPJ has more advantages concerning the flexibility of operation than other plasmas (Matsusaka, 2019). As the discharge extends to the outside of the electrode, the effect of the electrode itself on the starch can be ignored; and the contact between starch and high-voltage electrode is avoided, without the effect of the starch being discharged and thermal corrosion.

8.2.5 Corona Electrical Discharge Plasma (CEDP)

CEDP, a corona phenomenon, is a type of partial self-sustaining discharge of gas in erratic electric fields. Consequently, CEDP can generate denser and more energetic plasma locally than DBDP because of unequal discharge characteristics (acting on the pertinence starch areas to boost the efficiency of modification) (Laroque et al., 2022). The generation and operation of CEDP is relatively simple and low in cost, but its applicability is limited due to the small and uneven areas (Scholtz et al., 2015). The surface of starch particles may have serious damage under partial high-voltage discharge when are treated with CEDP.

8.3 STRUCTURAL CHANGE OF COLD PLASMA-TREATED STARCH (CPTS)

CPTS could modify starch via crosslinking/grafting, depolymerization, plasma etching, and the conversion of functional groups (Han et al., 2020) (Figure 8.4). The crosslinking is that a new C–O–C bond is produced when the reducing ends of two starch polymer chains (C–OH) break and water molecules are eliminated. When compared to other glucose molecule locations, the C-2 site is the most prone to cross-linking (80.6%) (Zou et al., 2004). The depolymerization is primarily caused by the ionization of water molecules. The high-energy particle ion bombardment leads to the depolymerization of starch chains and forms smaller fragments such as maltose, maltotriose, and maltotetraose (Thirumdas et al., 2017). The plasma input energy plays an important role in determining which is the dominant of the two competitive reactions between depolymerization and cross-linking (Wongsagonsup et al., 2014).

Plasma etching is also commonly used to modify the surface of starch granules and powders. Reactive chemicals scratch the starch surfaces during the plasma production process. Plasma can significantly modify the multi-scale structure of starch due to its great particle size penetration and active chemical fingerprints. Moreover, CP is typically combined with microwave, ultrasound, enzymatic hydrolysis, and other technologies to obtain superior modifying effects (Ge et al., 2021a, 2021b; Sun et al., 2022b).

The ordered architecture of starch from basic glucan units, glycosyl chains, and condensed structure to micron-sized granule shows an excellent processability with multi-level structural changes (Chi et al., 2021). The physical and chemical properties of the starch matrix change along with the multi-scale structures. Plasma mainly destroys the chain-length distribution (CLD) of starch, with the modification of crystalline lamella and amorphous region. For this reason, we can further illustrate the research progress of plasma in starch CLDs and lamellar structure and adjust the structure of starch. It achieves the ideal accurate function-oriented means of modification via changing the operating parameters of plasma.

8.3.1 Modification in Chain-length Distribution

Generally, when starch is only treated by plasma, the proportion of A chains and B1 chains increases, while the proportion of B2 chains and B3 chains decreases (Figure 8.3). This law seems to occur in five different starches (maize starch, corn starch, rice, lotus root, and sweet potato), which are treated by DBD-CP (Ge et al., 2021a, 2022; Liang et al., 2023; Sun et al., 2023). However, this phenomenon has not been completely verified in six types of starch from Gao et al. (2019, 2022) and Shen et al. (2021a) (with a result of A chains ↑, B1, B2, B3 chains ↓) and five types of starch from Ge et al. (2021b), Gupta et al. (2023), Shen et al. (2022a), Sun et al. (2021) (with a result of A chains ↑, B3 chains ↓, and B1, B2 chains show random changes). The alternation of CLD possibly shows a trend of A chains rising, B3 chains decreasing, and B1 chain always occupy the largest proportion, in addition to the CP-treated rice starch (with an opposite result of A chains and B1 chains ↓, B2 and B3 chains ↑) (Sun et al., 2022b). Ge et al. (2021a) observed that the overall CLD of the starch also moved toward the short-chain region (A, B1 chain ↑, B2, B3 chain ↓) when sweet potato starch was treated by DBDP combined with pullulanase. The same phenomenon occurs with the combined DBDP and α-amylase treatment of wheat starch (with a result of A chains ↑, B3 chains ↓, and B1, B2 chains show random changes) (Shen et al., 2022b). Therefore, the degree of

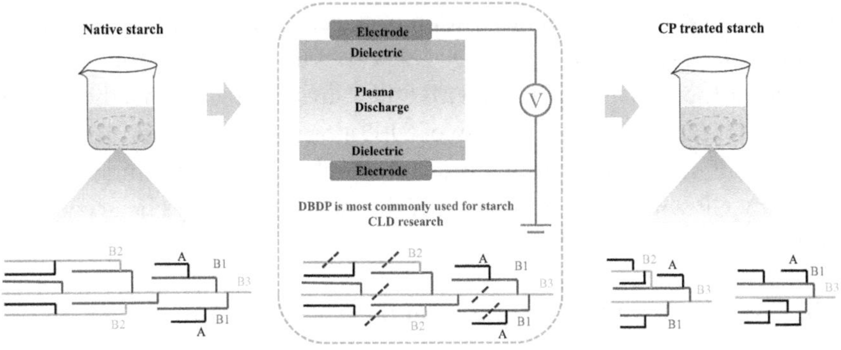

FIGURE 8.3 Diagram of starch chain-length distribution via dielectric barrier discharge plasma (DBDP).

depolymerization of amylopectin is likely greater than that of polymerization under reactive free radicals when starch-debranching enzymes are used to assist and enhance.

Non-thermal plasma readily targets starch CLDs and promotes the transformation from long chain to short chain. However, this rule may no longer apply when plasma is combined with other physical synergies. For red bean starch, for instance, DBDP in conjunction with dry heat treatment was used. The maximum DP reduced dramatically with an increase in the ultrashort chain of DP ≤ 6, even if the short chain increased in CLD within the range of A chains and B1 chains. B3 chains vanished immediately (Ge et al., 2021b). The proportion of A chains and B1 chains for rich starch treated with DBDP combined with microwave first decreased and then increased as the plasma treatment duration increased (Sun et al., 2022b). This demonstrated that the treatment period had an important effect on the amylopectin reorganization aside from treatment intensity and plasma type. Interestingly, Sun et al. (2021) found an opposite phenomenon, when the plasma (combined with extrusion) duration was increased, the B1 chains decreased and the A chains in starch initially increased and then decreased (albeit they were still higher than that of native starch). On the other hand, the proportion of B1 chains increased, the proportion of B3 chains decreased, and other chains exhibited erratic fluctuations when mung bean starch was treated by CP combined with ultrasound (Shen et al., 2021b). After synthesizing potato starch nanocrystals, Shen et al. (2022a) found that the CLD change rule was in line with that of common starch, i.e., the proportion of A chains and B1 chains increased while those of B2 and B3 chains dropped. The synergistic mode including non-thermal plasma and its treatment duration have a significant effect on the depolymerization and polymerization of starch amylopectin; however, the change rule of starch chains change rule varies depending on the source of starches. When plasma is mixed with other physical treatments, there is no clear rule (i.e. the basic mechanism of CLD alternation remains unknown). However, it seems that the majority of starch CLDs that remains unaffected by CP with other processing techniques may always be accounted for by B1 chains.

8.3.2 Modification in Amorphous and Crystal Structure

In addition to the starch's relative crystallinity, non-thermal plasma mostly impacts the amorphous portion of the material while having minimal effect on the A, B, or C types of crystal form (Banura et al., 2018;

Gupta et al., 2023; Sun et al., 2021). Notably, some A-type crystals in natural starch transformed to B-type crystals when Ariá starch was treated by DBDP. It was hypothesised that the shorter chains in the A-type crystal structures were more susceptible to being broken down into the B-type crystal structures by reactive water substances (Carvalho et al., 2021). Nevertheless, DBDP increased the crystallinity of banana starch to A-type crystals in cases where the crystals were C-type (more inclined toward B-type) (Yan et al., 2020). However, not all starch containing C-type crystals changed underwent CP treatment, as this varies depending on the plasma parameters and objects.

Obviously, the effects of CP on the amorphous regions of starch are evident, and CP-induced crystallinity alteration has recently drawn more attention. When the depolymerization of amylose chains is greater than their cross-linking due to the active substance of plasma, the amylose small fragments increase, the amorphous regions expand, and the relative crystallinity decreases. Guo et al. (2022) found that the relative crystallinity of modified starch decreased as the time extended when DBDP was used. This is consistent with the earlier findings that Shen et al. (2022a) and Sudheesh et al. (2019) reported. This interaction is reinforced by the rise in plasma power and action duration when the high-energy atoms and ions' average collision velocity and collision periods likewise increase. Compared with other physical and/or chemical methods, CP usually reduced the relative crystallinity of starch more quickly (Sun et al., 2022b; Ge et al., 2021a, 2021b).

Interestingly, compared to native starch, when starch molecular chains are cross-linked using plasma to form free radicals and high-energy electrons, the starch chains in the amorphous regions are more orderly in terms of cross-linking rearrangements and are closer to the crystalline regions via condensation and ether bond formation. Wu et al. (2018) found that the relative crystallinity of starch increased after utilizing CEDP to change banana starch. This also occurred in the study of Sun et al. (2021). Lii et al. (2002) found that crystallinity increased when cassava and waxy corn starches were treated with H, O, and NH_3 produced by low-pressure GDP. They formed more crystalline domains under appropriate regulation, in which led to the amplification of crystalline regions. Moreover, helium plasma was shown to have higher energy and be a more reactive species in terms of damage to the crystalline regions of the starch granules compared to argon plasma (Srangsomjit et al., 2022).

In addition, the crystalline and amorphous areas of starch are influenced by the natural crystal form (despite the fact that plasma dose not alter the crystal itself) and the manner of plasma generation also affect the crystalline and amorphous regions of starch. It has been proven that the branch point of B-type starch clusters is nearly in the amorphous region, which is vulnerable to plasma reaction (Jane et al., 1997). The crystallinity of the potato decreased when glutinous rice, corn, and potato were treated with CO_2-Ar RFP, the crystallinity of the corn and rice remained unaffected (Okyere et al., 2019). According to research, there is a difference in the water content between cereal starch (type A crystal; 8 crystalline water molecules) and tuber starch (type B crystal; comprising 36 crystalline water molecules). Additionally, excessive hydrolysis produces an active hydroxyl radical that reduces the crystallinity of potato starch. The DBDP treatment of potato starch showed the similar phenomenon (Ge et al., 2021a). Sifuentes-Nieves et al. (2020) used hexamethyldisiloxane (HMDSO) RFP to treat three types of corn starches with varying branching ratios. The results indicated that the content of amylopectin and the crystal shape had an impact on the area where the starch crystallized.

In summary, non-thermal plasma is a useful tool to control the amorphous region of starch. It also induces the formation of active water substances in starch crystals to change the degree of crystallinity. Though the crystallinity of starch always decreases, the potential recrystallization with increased crystals can be obtained largely depending on the sources of starch and the types and parameters of plasma.

8.3.3 Modification in Lamellar Structure

So far, there are only a few reports on the effects of plasma on the lamellar structure of starch. The damage of CP to the crystalline lamellae may be greater than that to the amorphous lamellae of starch (Figure 8.4). Wang et al. (2023) treated nut starch with DBDP and found that the crystalline layers of starch mildly decrease and the amorphous layers increase mildly. The same phenomenon occurred in the treated cassava starch with oxygen and helium GDP (Bie et al., 2016). They found that when the treatment time was more than 30 minutes, the semi-crystalline lamellae were only slightly affected, while the oxygen GDP induced more prominent destruction of the crystalline lamellae than the amorphous lamellae. On the contrary, when potato starches were treated with GDP, the average thickness of semi-crystalline lamellae was increased through an increase in the thickness of crystalline lamellae and an insignificant decrease in the thickness of amorphous lamellae (Zhang et al., 2015a). In particular, nitrogen is more easily affected by glow plasma than hydrogen, resulting

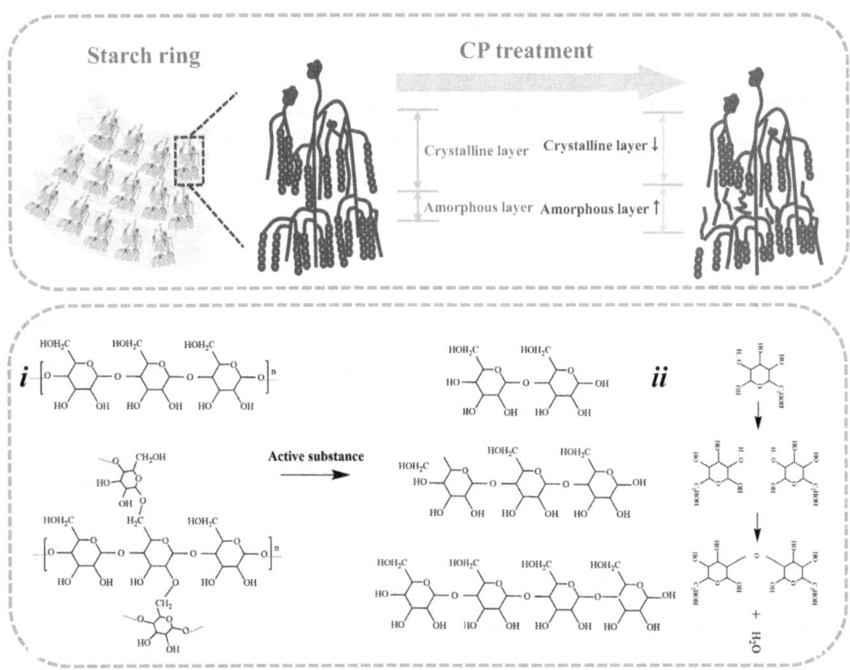

FIGURE 8.4 Diagram of the starch-layered structure via CP and the mechanism of starch modification by CP: (i) depolymerization and (ii) crosslinking/grafting.

in more active substances and effective changes in the characteristics of potato nanostructures.

Similarly, RFP treatment could enhance the semi-crystalline lamellae thickness of starch, and the intensity of the X-ray scattering peak of starch were changed but not significant (Zhou et al., 2023). When treating potato and corn starches with GDP, different treatment times had differential effects on the crystalline and amorphous regions. The degree of damage to the amorphous lamella was higher at 30 minutes, and the degree of damage to the crystalline lamella was higher at 60 minutes (Zhang et al., 2014). This means that the changes of electron density are different by comparing the crystal and the amorphous lamellae by GDP treatment at the same conditions.

According to the above information, the medium gas of plasma and the time of treatment significantly affect the lamellar structure of starch. The crystalline and amorphous regions structure of starches, as well as the semi-crystalline repetition distance and the average thickness of the semi-crystallization of starch lamellar structure, are all affected by the active compounds found in plasma. Currently, several one-dimensional theoretical models and small-angle X-ray scattering (SAXS) research are the primary sources of information about the lamellar properties of starch (see Chapter 1). Future research on lamellar structures may benefit from the development of tailored characterization techniques based on biomacromolecular approaches like cryo-electron microscopy (cryo-EM).

8.3.4 Modification in Granule Structure

Cracks, holes, etching, and micro-deposits appeared on the surface of starch particles due to the plasma corrosion while maintaining the integrity of the starch granules (Figure 8.5). When corn starch was treated via CP, it was found that the plasma could produce and enlarge the microporous channel extending from the surface to the core (Okyere et al., 2019). It improved of the plasma active substances penetration into the interior of the particles, which result in the depolymerization of starch with obvious fracture in some areas. The starch granules with holes or pins on the surface were more likely to generate further and larger holes, while small starch granules were more likely to aggregate than big ones after CP treatment (Gao et al., 2019). The surface damage of banana starch granules was greater with an increase in voltage intensity, and the shape was smaller and more irregular than native counterpart when treated with CEDP (Wu et al., 2018). The cornstarch granules showed uneven surfaces and deposits after exposure to the RFP treatment (Banura et al., 2018). Helium GDP was shown to impart

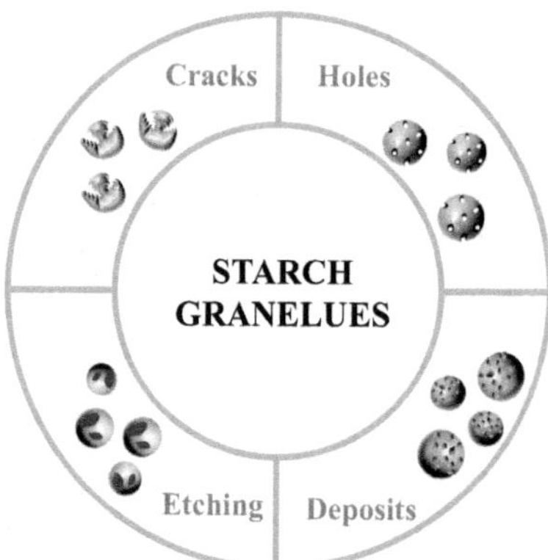

FIGURE 8.5 Diagram of starch granules via CP (four phenomena appeared on the surface of starch particles due to the plasma corrosion while maintaining the integrity of the starch granules).

visible corrosion damage to the granule surface of potato starch (Zhang et al., 2015b). However, the morphology of cassava starch granules was not significantly changed when treated by helium GDP, which may be related to the different starch sources (Bie et al., 2016). Furthermore, the surface of the starch particles could be caused serious damage due to the high-energy neutral nitrogen atoms produced by GDP, and large cracks and cavities were also formed. Serious damage to the surface of the starch particles was also caused by air plasma (Sudheesh et al., 2019).

Non-thermal plasma (such as DBD) combined with other processing methods (e.g., enzyme, dry heat, ultrasonic, microwave) can cause greater cracks and pores on the surface of starch particles than pure plasma (Ge et al., 2021a, 2021b; Shen et al., 2021b, 2022b). For example, when CP is combined with extrusion, the effects of thermomechanical energy and water flash during the extrusion process caused the starch particles to become irregular in shape with a rough surface (Sun et al., 2021). CP pretreatment created deeper pores on the porous starch, revealing that plasma etching made more surface available for the enzyme and promoting the formation of porous structure (Davoudi et al., 2022). However, the HMDSO coating could reduce the expansion of cavities and pores on the surface of the particles caused by plasma via covering the surface of the starch particles and increasing the aggregation between the starch particles (Sifuentes-Nieves et al., 2021).

When potato starch was made as nanoparticles (i.e., the starch particle size was minimized), most nanoparticles appeared in the form of aggregates. Nevertheless, CP treatment could help the nanoparticles become well dispersed and stay stable when prepared by vacuum CP combined with ultrasound (Chang et al., 2020). The nano-sized starch particles modified by pure oxygen and ammonia plasma made the nanoparticles have good dispersion, reduced the relative crystallinity, and accommodated more guest molecules (Chang et al., 2020). Similar results were reported by Chang et al. (2019) and Wang et al. (2022). More investigations on the surface plasma modification of starch at the nanoscale are needed with follow-up researches.

8.4 CPTS APPLICATIONS

8.4.1 Plasma for Starch Modification as food materials

The above-mentioned effects of CP on the hierarchical structure of starch are directly related to the uses of starch and starch-based foods. CP has been demonstrated to be an effective catalytic approach to promote starch modification. Ji et al. (2022) proposed a new modification approach for cassava starch succinylation prepared with atmospheric pressure plasma jets and esterified with octenyl succinic anhydride (OSA). To expand the reaction area between the OSA and starch, atmospheric pressure CP could degrade the starch surface and double helix structure. The succinylation efficiency of starch increased as the pretreatment time was increased. Similarly, Shen et al. (2021a) pretreated barley starch with DBDP before using it to esterify citrate. This pretreatment increased the substitution rate of citric acid, decreased starch enzyme digestibility, and increased digestion resistance.

Besides, CP is usually a promising technology for treating cereal grains and boosting seed germination and plant growth, thus improving food quality and security. DBDP is a new technique utilized in starchy foods that could prevent lipid oxidation and preserve the color of rice during storage (Liu et al., 2021). Also, DBDP might significantly reduce cooking time and hardness, and increase texture by producing rice surface fissures (Akasapu et al., 2020). Meanwhile, it was proven that DBDP could boost rice germination and seedling growth rate by etching the outermost seed coat of rice (Billah et al., 2021). This effect is also linked to numerous types of reactive species influenced by CP treatment, which is conducted on the epigenetic regulation of rice seeds (Suriyasak et al., 2021).

There are further investigations being started on the use of plasma as an environmentally friendly technique in starch-containing foods such as fruit by-products. For example, CP might improve the drying rate of jujube slices by etching huge voids on the surface of the jujube slices, which facilitates water migration (Bao et al., 2021). Similarly, plasma was found to have a substantial impact on the surface morphology of mango kernel starch, bringing changes in functional parameters such as decreased viscosity and increased hydration capacity (Kalaivendan et al., 2022).

8.4.2 Enhancement of Mechanical Strength for Packaging

CP as non-thermal treatment is appropriate for the modification of thermally sensitive starch materials, and CP-treated films made from polysaccharide has been focused with the advanced surface modification (Desmet et al., 2009). There are two basic strategies of processing for preparing starch-based film relating to CP. One is using CP to modify starch for further preparing film, and another method is using CP to modify starch film directly. The positive action of CP to starch-based films has gradually been demonstrated with their properties of water vapor permeability, hydrophilicity, and tensile characteristics improved for using as packaging materials.

From a perspective of structure, the starch film treated with CP was smooth and homogeneous (Guo et al., 2022). HMDSO RFP was used to modify starch for preparing film from amylose (Sifuentes-Nieves et al., 2020). Amylose molecules quickly flowed out of cracks and pores during the film-forming process during CP treating, with the increase of the solubility, and viscosity of product. HMDSO was able to fully coat the surface of starch particles and function as a barrier to block amylopectin outflow. Starch-based films treated by HMDSO RFP might be soften with the improvement of hydrophobicity as reported by Hernandez-Perez et al. (2021). Furthermore, CP can be utilized as "adhesive" to reinforce the interfacial structure by facilitate the adhesion of starch and polyethylene terephthalate (Wiącek et al., 2017). DBDP was used to treat multilayer films with cassava starch, polycaprolactone (PCL), and polylactic acid (PLA) (Heidemann et al., 2019), with starch adhesion and delamination resistance in PCL/PLA films were strengthened.

The functional properties of starch film are important for utilization, and Goiana et al. (2021) found the enhancement of the hydrophobicity, tensile strength, and stiffness of the CP-treated films. While Sudheesh et al. (2020) reported that when starch-based films were treated with different intensities of CP, their properties all reduced dramatically including water vapor permeability, oxygen permeability, solubility, etc. In order to improve the composite film's tensile characteristics and hydrophilicity, Sheikhi et al. (2021) treated the starch/chitosan film in air and argon plasma. Also, Florez et al. (2019) treated polyhydroxybutyrate (PHB) by using atmosphere and sulfur hexafluoride plasma and added PHB as a reinforcing agent to thermoplastic starch. This approach increased the mechanical characteristics, tensile strength, and elastic modulus of starch sheet, as well as its thermal stability.

REFERENCE, BIBLIOGRAPHY OR WORKS CITED

Akasapu, K., Ojah, N., Gupta, A. K., Choudhury, A. J., & Mishra, P. 2020. An innovative approach for iron fortification of rice using cold plasma. *Food Research International* 136: 109599.

Banura, S., Thirumdas, R., Kaur, A., Deshmukh, R. R., & Annapure, U. S. 2018. Modification of starch using low pressure radio frequency air plasma. *LWT* 89: 719–724.

Bao, T., Hao, X., Shishir, M. R. I., Karim, N., & Chen, W. 2021. Cold plasma: An emerging pretreatment technology for the drying of jujube slices. *Food Chemistry* 337: 127783.

Bie, P., Li, X., Xie, F., Chen, L., Zhang, B., & Li, L. 2016. Supramolecular structure and thermal behavior of cassava starch treated by oxygen and helium glow-plasmas. *Innovative Food Science & Emerging Technologies* 34: 336–343.

Billah, M., Karmakar, S., Mina, F. B., Haque, M. N., Rashid, M. M., Hasan, M. F., Acharjee, U. K., & Talukder, M. R. 2021. Investigation of mechanisms involved in seed germination enhancement, enzymatic activity and seedling growth of rice (*Oryza sativa* L.) using LPDBD (Ar+Air) plasma. *Archives of Biochemistry and Biophysics* 698: 108726.

Carvalho, A. P. M. G., Barros, D. R., Da Silva, L. S., Sanches, E. A., Pinto, C. D. C., de Souza, S. M., Clerici, M. T. P. S., Rodrigues, S., Fernandes, F. A. N., & Campelo, P. H. 2021. Dielectric barrier atmospheric cold plasma applied to the modification of Ariá (*Goeppertia allouia*) starch: Effect of plasma generation voltage. *International Journal of Biological Macromolecules* 182: 1618–1627.

Chang, R., Ji, N., Li, M., Qiu, L., Sun, C., Bian, X., Qiu, H., Xiong, L., & Sun, Q. 2019. Green preparation and characterization of starch nanoparticles using a vacuum cold plasma process combined with ultrasonication treatment. *Ultrasonics Sonochemistry* 58: 104660.

Chang, R., Lu, H., Tian, Y., Li, H., Wang, J., & Jin, Z. 2020. Structural modification and functional improvement of starch nanoparticles using vacuum cold plasma. *International Journal of Biological Macromolecules* 145: 197–206.

Chi, C., Li, X., Huang, S., Chen, L., Zhang, Y., Li, L., & Miao, S. 2021. Basic principles in starch multi-scale structuration to mitigate digestibility: A review. *Trends in Food Science & Technology* 109: 154–168.

Chizoba Ekezie, F., Sun, D., & Cheng, J. 2017. A review on recent advances in cold plasma technology for the food industry: Current applications and future trends. *Trends in Food Science & Technology* 69: 46–58.

Davoudi, Z., Azizi, M. H., & Barzegar, M. (2022). Porous corn starch obtained from combined cold plasma and enzymatic hydrolysis: Microstructure and physicochemical properties. *International Journal of Biological Macromolecules* 223: 790–797.

Desmet, T., Morent, R., De Geyter, N., Leys, C., Schacht, E., & Dubruel, P. 2009. Nonthermal plasma technology as a versatile strategy for polymeric biomaterials surface modification: A review. *Biomacromolecules* 10: 9.

Dhakal, O. B., Dahal, R., Acharya, T. R., Lamichhane, P., Gautam, S., Lama, B., Khanal, R., Kaushik, N. K., Choi, E. H., & Chalise, R. 2023. Effects of spark dielectric barrier discharge plasma on water sterilization and seed germination. *Current Applied Physics* 54: 49–58.

Florez, J. P., Fazeli, M., & Simão, R. A. 2019. Preparation and characterization of thermoplastic starch composite reinforced by plasma-treated poly (hydroxybutyrate) PHB. *International Journal of Biological Macromolecules* 123: 609–621.

Gao, S., Liu, H., Sun, L., Liu, N., Wang, J., Huang, Y., Wang, F., Cao, J., Fan, R., Zhang, X., & Wang, M. 2019. The effects of dielectric barrier discharge plasma on physicochemical and digestion properties of starch. *International Journal of Biological Macromolecules* 138: 819–830.

Gao, S., Zhang, H., Pei, J., Liu, H., Lu, M., Chen, J., & Wang, M. 2022. High-voltage and short-time dielectric barrier discharge plasma treatment affects structural and digestive properties of Tartary buckwheat starch. *International Journal of Biological Macromolecules* 213: 268–278.

Ge, X., Guo, Y., Zhao, J., Zhao, J., Shen, H., & Yan, W. 2022. Dielectric barrier discharge cold plasma combined with cross-linking: An innovative way to modify the multi-scale structure and physicochemical properties of corn starch. *International Journal of Biological Macromolecules* 215: 465–476.

Ge, X., Shen, H., Su, C., Zhang, B., Zhang, Q., Jiang, H., Yuan, L., Yu, X., & Li, W. 2021a. Pullulanase modification of granular sweet potato starch: Assistant effect of dielectric barrier discharge plasma on multi-scale structure, physicochemical properties. *Carbohydrate Polymers* 272: 118481.

Ge, X., Shen, H., Su, C., Zhang, B., Zhang, Q., Jiang, H., & Li, W. 2021b. The improving effects of cold plasma on multi-scale structure, physicochemical and digestive properties of dry heated red adzuki bean starch. *Food Chemistry* 349: 129159.

Goiana, M. L., de Brito, E. S., Alves Filho, E. G., Miguel, E. D. C., Fernandes, F. A. N., Azeredo, H. M. C. D., & Rosa, M. D. F. 2021. Corn starch-based films treated by dielectric barrier discharge plasma. *International Journal of Biological Macromolecules* 183: 2009–2016.

Guo, Z., Gou, Q., Yang, L., Yu, Q., & Han, L. 2022. Dielectric barrier discharge plasma: A green method to change structure of potato starch and improve physicochemical properties of potato starch films. *Food Chemistry* 370: 130992.

Gupta, R. K., Guha, P., & Srivastav, P. P. 2023. Effect of high voltage dielectric barrier discharge (DBD) atmospheric cold plasma treatment on physicochemical and functional properties of taro (*Colocasia esculenta*) starch. *International Journal of Biological Macromolecules* 253: 126772.

Han, Z., Shi, R., & Sun, D. 2020. Effects of novel physical processing techniques on the multi-structures of starch. *Trends in Food Science & Technology* 97: 126-135.

Heidemann, H. M., Dotto, M. E. R., Laurindo, J. B., Carciofi, B. A. M., & Costa, C. 2019. Cold plasma treatment to improve the adhesion of cassava starch films onto PCL and PLA surface. *Colloids and Surfaces A: Physicochemical and Engineering Aspects* 580: 123739.

Hernandez-Perez, P., Flores-Silva, P. C., Velazquez, G., Morales-Sanchez, E., Rodríguez-Fernández, O., Hernández-Hernández, E., Mendez-Montealvo, G., & Sifuentes-Nieves, I. 2021. Rheological performance of film-forming solutions made from plasma-modified starches with different amylose/amylopectin content. *Carbohydrate Polymers* 255: 117349.

Jane, J., Wong, K., & McPherson, A. E. 1997. Branch-structure difference in starches of A- and B-type X-ray patterns revealed by their Naegeli dextrins. *Carbohydrate Research* 300 (3): 219–227.

Jha, N., Ryu, J. J., Choi, E. H., Kaushik, N. K., & Lillig, C. H. 2017. Generation and role of reactive oxygen and nitrogen species induced by plasma, lasers, chemical agents, and other systems in dentistry. *Oxidative Medicine and Cellular Longevity* 2017: 7542540.

Ji, S., Xu, T., Huang, W., Gao, S., Zhong, Y., Yang, X., Ahmed Hassan, M., & Lu, B. 2022. Atmospheric pressure plasma jet pretreatment to facilitate cassava starch modification with octenyl succinic anhydride. *Food Chemistry* 370: 130922.

Kalaivendan, R. G. T., Mishra, A., Eazhumalai, G., & Annapure, U. S. 2022. Effect of atmospheric pressure non-thermal pin to plate plasma on the functional, rheological, thermal, and morphological properties of mango seed kernel starch. *International Journal of Biological Macromolecules* 196: 63–71.

Kavitha, E. R., Meiyazhagan, S., Yugeswaran, S., Balraju, P., & Suresh, K. 2021. Electrochemical prospects and potential of hausmannite Mn_3O_4 nanoparticles synthesized through microplasma discharge for supercapacitor applications. *International Journal of Energy Research* 45 (5): 7038–7056.

Kumari, B., & Sit, N. 2023. Comprehensive review on single and dual modification of starch: Methods, properties and applications. *International Journal of Biological Macromolecules* 253: 126952.

Laroque, D. A., Seó, S. T., Valencia, G. A., Laurindo, J. B., & Carciofi, B. A. M. 2022. Cold plasma in food processing: Design, mechanisms, and application. *Journal of Food Engineering* 312: 110748.

Li, J., Gao, H., Lan, C., Nie, L., Liu, D., Lu, X., & Ken Ostrikov, K. 2023. Plasma air filtration system for intercepting and inactivation of pathogenic microbial aerosols. *Journal of Environmental Chemical Engineering* 11 (5): 110728.

Li, S., Dang, X., Yu, X., Abbas, G., Zhang, Q., & Cao, L. 2020. The application of dielectric barrier discharge non-thermal plasma in VOCs abatement: A review. *Chemical Engineering Journal* 388: 124275.

Liang, W., Zhang, Q., Duan, H., Zhou, S., Zhou, Y., Li, W., & Yan, W. 2023. Understanding $CaCl_2$ induces surface gelatinization to promote cold plasma modified maize starch: Structure-effect relations. *Carbohydrate Polymers* 320: 121200.

Lii, C., Liao, C., Stobinski, L., & Tomasik, P. 2002. Effects of hydrogen, oxygen, and ammonia low-pressure glow plasma on granular starches. *Carbohydrate Polymers* 49 (4): 449–456.

Liu, P., Hou, M., & Wang, C. 2022. Interaction of glow discharge plasma with starch. *Journal of Chinese Institute of Food Science and Technology* 22 (6): 344-357.

Liu, Q., Wu, H., Luo, J., Liu, J., Zhao, S., Hu, Q., & Ding, C. 2021. Effect of dielectric barrier discharge cold plasma treatments on flavor fingerprints of brown rice. *Food Chemistry* 352: 129402.

Matsusaka, S. 2019. Control of particle charge by atmospheric pressure plasma jet (APPJ): A review. *Advanced Powder Technology* 30 (12): 2851–2858.

Meiyazhagan, S., Yugeswaran, S., Ananthapadmanabhan, P. V., & Suresh, K. 2020. Process and kinetics of dye degradation using microplasma and its feasibility in textile effluent detoxification. *Journal of Water Process Engineering* 37: 101519.

Moradi, D., Ramezan, Y., Eskandari, S., Mirsaeedghazi, H., & Javanmard Dakheli, M. 2023. Plasma-treated LDPE film incorporated with onion and potato peel extract – A food packaging for shelf life extension on chicken thigh. *Food Packaging and Shelf Life* 35: 101012.

Moreau, M., Orange, N., & Feuilloley, M. G. J. 2008. Non-thermal plasma technologies: New tools for bio-decontamination. *Biotechnology Advances* 26 (6): 610–617.

Okyere, A. Y., Bertoft, E., & Annor, G. A. 2019. Modification of cereal and tuber waxy starches with radio frequency cold plasma and its effects on waxy starch properties. *Carbohydrate Polymers* 223: 115075.

Okyere, A. Y., Rajendran, S., & Annor, G. A. 2022. Cold plasma technologies: Their effect on starch properties and industrial scale-up for starch modification. *Current Research in Food Science* 5: 451–463.

Saremnezhad, S., Soltani, M., Faraji, A., & Hayaloglu, A. A. 2021. Chemical changes of food constituents during cold plasma processing: A review. *Food Research International* 147: 110552.

Scholtz, V., Pazlarova, J., Souskova, H., Khun, J., & Julak, J. 2015. Nonthermal plasma - A tool for decontamination and disinfection. *Biotechnology Advances* 33 (6, Part 2): 1108–1119.

Sheikhi, Z., Mirmoghtadaie, L., Abdolmaleki, K., Khani, M. R., Farhoodi, M., Moradi, E., Shokri, B., & Shojaee-Aliabadi, S. 2021. Characterization of physicochemical and antimicrobial properties of plasma-treated starch/chitosan composite film. *Packaging Technology and Science* 34 (7): 385–392.

Shen, H., Ge, X., Zhang, B., Su, C., Zhang, Q., Jiang, H., Zhang, G., Yuan, L., Yu, X., & Li, W. 2022a. Preparing potato starch nanocrystals assisted by dielectric barrier discharge plasma and its multiscale structure, physicochemical and rheological properties. *Food Chemistry* 372: 131240.

Shen, H., Ge, X., Zhang, B., Su, C., Zhang, Q., Jiang, H., Zhang, G., & Li, W. 2021a. Understanding the multi-scale structure, physicochemical properties and in vitro digestibility of citrate naked barley starch induced by non-thermal plasma. *Food & Function* 12 (17): 8169–8180.

Shen, H., Ge, X., Zhang, Q., Zhang, X., Lu, Y., Jiang, H., Zhang, G., & Li, W. 2022b. Dielectric barrier discharge plasma improved the fine structure, physicochemical properties and digestibility of α-amylase enzymatic wheat starch. *Innovative Food Science & Emerging Technologies* 78: 102991.

Shen, H., Guo, Y., Zhao, J., Zhao, J., Ge, X., Zhang, Q., & Yan, W. 2021b. The multiscale structure and physicochemical properties of mung bean starch modified by ultrasound combined with plasma treatment. *International Journal of Biological Macromolecules* 191: 821–831.

Sifuentes-Nieves, I., Mendez-Montealvo, G., Flores-Silva, P. C., Nieto-Pérez, M., Neira-Velazquez, G., Rodriguez-Fernandez, O., Hernández-Hernández, E., & Velazquez, G. 2021. Dielectric barrier discharge and radio-frequency plasma effect on structural properties of starches with different amylose content. *Innovative Food Science & Emerging Technologies* 68: 102630.

Sifuentes-Nieves, I., Velazquez, G., Flores-Silva, P. C., Hernández-Hernández, E., Neira-Velázquez, G., Gallardo-Vega, C., & Mendez-Montealvo, G. 2020. HMDSO plasma treatment as alternative to modify structural properties of granular starch. *International Journal of Biological Macromolecules* 144: 682–689.

Sigeneger, F., Schäfer, J., Weltmann, K., Foest, R., & Loffhagen, D. 2017. Modeling of a non-thermal RF plasma jet at atmospheric pressure. *Plasma Processes and Polymers* 14 (4–5): 1600112.

Šimončicová, J., Kryštofová, S., Medvecká, V., Ďurišová, K., & Kaliňáková, B. 2019. Technical applications of plasma treatments: Current state and perspectives. *Applied Microbiology and Biotechnology* 103 (13): 5117–5129.

Srangsomjit, N., Bovornratanaraks, T., Chotineeranat, S., & Anuntagool, J. 2022. Solid-state modification of tapioca starch using atmospheric nonthermal dielectric barrier discharge argon and helium plasma. *Food Research International* 162: 111961.

Sreedevi, P. R., & Suresh, K. 2023. Cold atmospheric plasma mediated cell membrane permeation and gene delivery-empirical interventions and pertinence. *Advances in Colloid and Interface Science* 320: 102989.

Sudheesh, C., Sunooj, K. V., Sasidharan, A., Sabu, S., Basheer, A., Navaf, M., Raghavender, C., Sinha, S. K., & George, J. 2020. Energetic neutral N2 atoms treatment on the kithul (*Caryota urens*) starch biodegradable film: Physicochemical characterization. *Food Hydrocolloids* 103: 105650.

Sudheesh, C., Sunooj, K. V., Sinha, S. K., George, J., Kumar, S., Murugesan, P., Arumugam, S., Ashwath Kumar, K., & Sajeev Kumar, V. A. 2019. Impact of energetic neutral nitrogen atoms created by glow discharge air plasma on the physico-chemical and rheological properties of kithul starch. *Food Chemistry* 294: 194–202.

Sun, X., Saleh, A. S. M., Lu, Y., Sun, Z., Zhang, X., Ge, X., Shen, H., Yu, X., & Li, W. 2022a. Effects of ultra-high pressure combined with cold plasma on structural, physicochemical, and digestive properties of proso millet starch. *International Journal of Biological Macromolecules* 212: 146–154.

Sun, X., Saleh, A. S. M., Sun, Z., Ge, X., Shen, H., Zhang, Q., Yu, X., Yuan, L., & Li, W. 2022b. Modification of multi-scale structure, physicochemical properties, and digestibility of rice starch via microwave and cold plasma treatments. *LWT* 153: 112483.

Sun, X., Sun, Z., Guo, Y., Zhao, J., Zhao, J., Ge, X., Shen, H., Zhang, Q., & Yan, W. 2021. Effect of twin-xuscrew extrusion combined with cold plasma on multi-scale structure, physicochemical properties, and digestibility of potato starches. *Innovative Food Science & Emerging Technologies* 74: 102855.

Sun, X., Sun, Z., Saleh, A. S. M., Lu, Y., Zhang, X., Ge, X., Shen, H., Yu, X., & Li, W. 2023. Effects of various microwave intensities collaborated with different cold plasma duration time on structural, physicochemical, and digestive properties of lotus root starch. *Food Chemistry* 405:134837.

Suriyasak, C., Hatanaka, K., Tanaka, H., Okumura, T., Yamashita, D., Attri, P., Koga, K., Shiratani, M., Hamaoka, N., & Ishibashi, Y. 2021. Alterations of DNA methylation caused by cold plasma treatment restore delayed germination of heat-stressed rice (*Oryza sativa* L.) seeds. *ACS Agricultural Science & Technology* 1 (1): 5-10.

Thirumdas, R., Kadam, D., & Annapure, U. S. 2017. Cold plasma: An alternative technology for the starch modification. *Food Biophysics* 12 (1): 129–139.

Thirumdas, R., Saragapani, C., Ajinkya, M. T., Deshmukh, R. R., & Annapure, U. S. 2016. Influence of low pressure cold plasma on cooking and textural properties of brown rice. *Innovative Food Science & Emerging Technologies* 37: 53–60.

Wagatsuma, K. 2020. Characteristics of the transient signal from pulsed glow discharge plasma for atomic emission analysis. *Applied Spectroscopy Reviews* 55 (1): 76–86.

Wang, G., Li, C., Zhang, X., Wang, Q., Cao, R., Liu, X., Yang, X., & Sun, L. 2023. The changed multiscale structures of tight nut (*Cyperus esculentus*) starch decide its modified physicochemical properties: The effects of non-thermal and thermal treatments. *International Journal of Biological Macromolecules* 253: 126626.

Wang, J., Yu, Y., Zhang, Z., Wu, W., Sun, P., Cai, M., & Yang, K. 2022. Formation of sweet potato starch nanoparticles by ultrasonic-assisted nanoprecipitation: Effect of cold plasma treatment. *Frontiers in Bioengineering and Biotechnology* 10: 986033.

Wang, X., Zhou, M., & Jin, X. 2012. Application of glow discharge plasma for wastewater treatment. *Electrochimica Acta* 83: 501–512.

Wiącek, A. E., Jurak, M., Gozdecka, A., & Worzakowska, M. 2017. Interfacial properties of PET and PET/starch polymers developed by air plasma processing. *Colloids and Surfaces A: Physicochemical and Engineering Aspects* 532: 323–331.

Wongsagonsup, R., Deeyai, P., Chaiwat, W., Horrungsiwat, S., Leejariensuk, K., Suphantharika, M., Fuongfuchat, A., & Dangtip, S. 2014. Modification of tapioca starch by non-chemical route using jet atmospheric argon plasma. *Carbohydrate Polymers* 102: 790–798.

Wu, T., Sun, N., & Chau, C. 2018. Application of corona electrical discharge plasma on modifying the physicochemical properties of banana starch indigenous to Taiwan. *Journal of Food and Drug Analysis* 26 (1): 244–251.

Xiao, G., Xu, W., Wu, R., Ni, M., Du, C., Gao, X., Luo, Z., & Cen, K. 2014. Non-thermal plasmas for VOCs abatement. *Plasma Chemistry and Plasma Processing* 34 (5): 1033–1065.

Yan, S., Chen, G., Hou, Y., & Chen, Y. 2020. Improved solubility of banana starch by dielectric barrier discharge plasma treatment. *International Journal of Food Science & Technology* 55 (2): 641–648.

Zhang, B., Chen, L., Li, X., Li, L., & Zhang, H. 2015a. Understanding the multi-scale structure and functional properties of starch modulated by glow-plasma: A structure-functionality relationship. *Food Hydrocolloids* 50: 228–236.

Zhang, B., Chen, L., Li, X., Li, L., & Zhang, H. 2015b. Understanding the multi-scale structure and functional properties of starch modulated by glow-plasma: A structure-functionality relationship. *Food Hydrocolloids* 50: 228–236.

Zhang, B., Xiong, S., Li, X., Li, L., Xie, F., & Chen, L. 2014. Effect of oxygen glow plasma on supramolecular and molecular structures of starch and related mechanism. *Food Hydrocolloids* 37: 69–76.

Zhang, K., Zhang, Z., Zhao, M., Milosavljević, V., Cullen, P. J., Scally, L., Sun, D., & Tiwari, B. K. 2022. Low-pressure plasma modification of the rheological properties of tapioca starch. *Food Hydrocolloids* 125: 107380.

Zhong, H., Shneider, M. N., Mokrov, M. S., & Ju, Y. 2019. Thermal-chemical instability of weakly ionized plasma in a reactive flow. *Journal of Physics D: Applied Physics* 52 (48): 484001.

Zhou, D., Yang, G., Tian, Y., Kang, J., & Wang, S. 2023. Different effects of radio frequency and heat block treatments on multi-scale structure and pasting properties of maize, potato, and pea starches. *Food Hydrocolloids* 136: 108306.

Zou, J., Liu, C., & Eliasson, B. 2004. Modification of starch by glow discharge plasma. *Carbohydrate Polymers* 55 (1): 23–26.

CHAPTER 9

Physical Preparation, Morphology, and Application of Nanosized Starch Materials

Siyu Yao and Enbo Xu

9.1 INTRODUCTION

Native starch is seldom used as functional materials or edible products directly due to a variety of limitations including low cold-water solubility, poor gel stability, dispersibility, and tolerance to processing or *in-vivo* digestion conditions (Dong et al., 2021; Guo et al., 2015). Thanks to nanotechnology as one of the best modification methods for material's structure (Campelo, Sant'Ana, & Pedrosa Silva Clerici, 2020; Chu et al., 2018; Wang & Lin et al., 2021; Yan et al., 2021), starch can be processed to a nanosized target (so-called nano-starch or starch nanoparticle, which is firstly found and prepared in this century, Figure 9.1) that attracts great attention owing to its high biocompatibility, easy availability, low sensitization, and the structure changes relating to small size and diversity morphologies for applications in the fields of food, medicine, and cosmetics

DOI: 10.1201/9781003493594-9

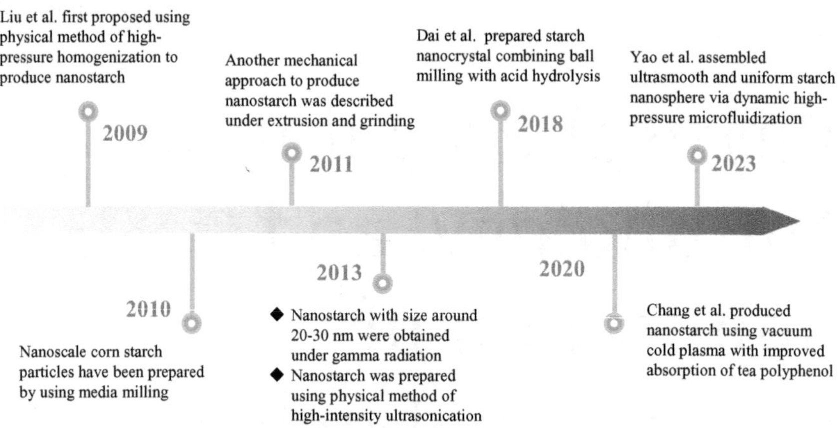

FIGURE 9.1 Timeline of important events held in the field of nano-starch preparation via the physical method.

(Bel Haaj & Thielemans et al., 2016; Yu et al., 2018; Zhu, 2017). Therefore, a major focus of research is the fabrication of nano-starch materials via green and high-productivity method.

Nano-starch is usually prepared by biological methods including enzymatic hydrolysis such as pullulanase (Wang & Chang et al., 2021), α-amylase (Qiu et al., 2019), and β-amylase (Chang et al., 2019); chemical methods such as oxidation (Yun et al., 2019; Zuo et al., 2017), etherification (Ye et al., 2019), and acid hydrolysis (Miskeen, An, & Kim, 2021; Qiu et al., 2019); and physical method such as vacuum cold plasma (Chang et al., 2020), ultrasonication (Chang & Yan et al., 2017), dynamic high-pressure microfluidization (DHPM) (Yao et al., 2023), gamma irradiation (Lamanna et al., 2013), and milling (Lu, Xiao, & Huang, 2018). Among them, physical methods are greener and more sustainable with high-efficiency production (Figure 9.1). In this chapter, we mainly summarize the physical techniques including high-pressure action, ultrasonication, gamma irradiation, and milling to obtain starch nanoparticles and systematically elucidate the preparation principles. For example, Ahmad and Lim et al. (2020) prepared nano-starch formed by high-pressure action with shear force and cavitation, with an improved water vapour permeability. Apostolidis and Mandala (2020) assembled starch nanoparticles by adjusting the pressure and frequencies. Wang and Lin et al. (2021) used high-pressure-assisted enzymatic hydrolysis to decrease the size of starch nanoparticles. However, the use of traditional static high-pressure process to improve properties and structures of nano-starch is to some extent limited, and dynamic high-pressure action with stronger forces (see Section 9.2) is needed to efficiently modify starch nano-structure and properties, and even may induce the molecular chemical changes (Zhu, 2021). Another case is an ultrasonic method which shows the cavitation effect to generate strong shear forces and shock waves near the starch chains, breaking the C–C bonds of macromolecules, forming more free radicals, and increasing the free energy. This helps to increase the dispersion behaviour of nano-starch after ultrasound (Ruan et al., 2022) and to achieve efficient encapsulation of bioactives (details in Section 9.4). Also, gamma radiation is an advantageous alternative to preparing starch nanoparticles due to its simple methodology. Gamma radiation generates free radicals on starch molecules, hydrolysing chemical bonds, thereby splitting large molecules of starch into nano-fragments (Yu & Wang, 2007). As an environmentally friendly physical method, milling treatment has been used to modify starch so as to produce micro- and nano-starch particles. The grinding ball will continue to move and rotate, producing kinetic energy, resulting in the decreased size of the sample (Lu et al., 2018).

Obviously, physical treatments are promising techniques to prepare nanosized starch with improved properties including particle size, morphology, dispersion ability, crystalline, and other functional characteristics. The outstanding advantages of nano-starch as green and edible materials include large specific surface area, good biocompatibility, non-toxicity, etc. (Qiu et al., 2020). Besides, under physical methods assisted with/without other chemical/biological methods, nano-starch can also be tailored into different shapes such as spherical, rod-shaped, worm-like, vesicular, micellar, lamellar, network form, and to some extent irregular morphologies in Section 9.3 (Yao et al., 2024). Furthermore, we design this chapter on advanced nano-starch materials prepared with the assistance of physical actions, which is expected to provide useful information on the applications of emulsion and delivery systems, as well as other potential fields.

9.2 PHYSICAL PREPARATION CHARACTERISTICS OF NANO-STARCH

9.2.1 High-Pressure Action

High-pressure homogenization (HPH) with shear force and cavitation has been commonly used to prepare nano-starch. Ahmad and Lim et al. (2020) developed nano-starch with improved water vapour permeability by using the HPH method. Nano-starch could be prepared with a relatively small particle size of 540 nm by adjusting the pressure and frequencies of HPH (Apostolidis & Mandala, 2020). Some researchers further used HPH-assisted enzymatic hydrolysis to decrease the size of nano-starch from 507.4 to 87.1 nm (Wang, Lin, et al., 2021). However, the function of traditional HPH to improve properties and structures of nano-starch is limited with a low production efficiency.

DHPM, as a novel physical method updated from HPH (Figure 9.2), involves combination forces including high-pressure effect, high-speed

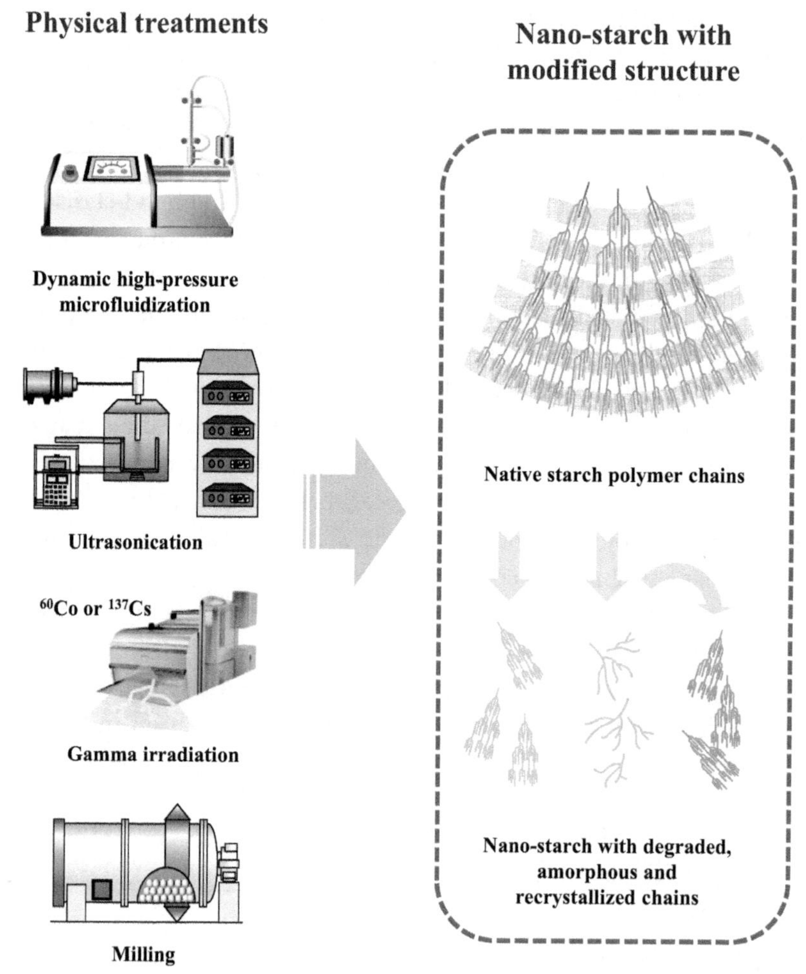

FIGURE 9.2 The effect of physical treatments (such as high-pressure microfluidization, ultrasonication, gamma irradiation, and milling) on starch structure for nanosized formation.

impact, high-frequency vibration, instantaneous pressure drops, strong shearing, and hydrodynamic cavitation to destroy the structures of starch granules (Wu et al., 2021). The diamond interaction chamber equipped with Y- or Z-shaped is the key component of a dynamic high-pressure homogenizer. In the process, the pump sends the starch suspension to the instrument, and the fluids are divided into multiple streams that have a strong high-speed impact in the diamond interaction chamber. The high-impact energy generated by the collision of micro-flows can damage the condensed structure of starch, thereby producing nanoparticles (Alkanawati et al., 2018; Ozturk & Turasan, 2021).

DHPM alters the surface and crystal structure of starch, and exposes large active hydroxyl groups, which is beneficial for nano-starch formation. DHPM treatment tends to significantly reduce particle size through molecular breakage (Bitik, Sumnu, & Oztop, 2019; Chen et al., 2021; He et al., 2020; Mei et al., 2020). Starch large molecules are preferentially disrupted by mechanical forces. Previous report confirmed that the presence of α-1,4-glycosidic bonds instead of α-1,6-glycosidic bond is prone to fracture under mechanical impact, showing a midpoint scission mechanism. Yao et al. (2023) used DHPM to prepare nano-starch (~67 nm) from gelatinized starch solution and found that the longer main (DP 100–10,000) and sub-main chains (i.e., B3 chain and B2 chain) can be firstly attacked, followed by side chain (i.e., B1 chain), which is in an inside-out mode. Also, DHPM changes the properties of starch, for example, Bitik et al. (2019) used DHPM to treat lentil and chickpea starch, showing decreased gelatinization temperatures and apparent viscosities. Kasemwong et al. (2011) found that the gelatinization enthalpy of cassava starch was pronounced decrease from 12.0 J/g (that of native counterpart) to 3.0 J/g by DHPM.

9.2.2 Ultrasonication

Ultrasonication is an effective approach of preparing nano-starch. During the ultrasonication process (Wang et al., 2023) (Figure 9.2), high energy is released through ultrasonic cavitation to generate strong shear forces and shock waves near the starch chain, where the microbubbles collapse in the form of sound and the energy is converted into high temperature and pressure to attack starch chains. The covalent bonds and chains of starch are broken and the size of starch particle is decreased to form a nanosized structure (Marta et al., 2022). Bel Haaj et al. (2013) treated starch suspension with ultrasound for 75 min, resulting in the formation of nano-starch (size between 30 and 100 nm). Ultrasonication is generally used along with other methods to degrade starch; for instance, Kim et al. (2013) found that acid hydrolysis at a low temperature of 40°C for 6 days followed by ultrasonication could effectively prepare globular nano-starch with size ranging from 50 to 90 nm. Bajer et al. (2023) also combined acid hydrolysis and sonication for nano-starch formation (5–500 nm), and explored the effect of amylose content on its property.

Compared to waxy maize starch, amylose-rich corn starch was more susceptible to sonication and acid method, and its crystallinity was significantly decreased. Ultrasonication has also been reported to break down the aggregation of the starch nanoparticles. Ruan et al. (2022) used rapid ultrasonication with periodic cavitation to disintegrate nano-starch aggregate into microparticles. However, when the ultrasonication was treated over time, the nano-starch would be reaggregated and was not conductive to the formation of uniform and dispersed samples.

9.2.3 Gamma Irradiation

Gamma irradiation (γ-radiation) has been usually used for polymer materials through grafting, cross-linking, and degradation (Figure 9.2) (Koubaa et al., 2016). Gamma irradiation has been reported as a convenient method for the size reduction of starch, which can break large molecules into smaller fragments and cleave glycosidic linkages (Singh et al., 2011). This action may generate active free radicals, which can break chemical bonds of starch molecules and then split macromolecular starch into smaller dextrin fragments in the amorphous regions. It is expected that α-d-(1–4)-glycoside linkages are the most susceptible to γ-radiation (Yu & Wang, 2007). (1) The excited fragment (RH*) of starch molecules generates due to irradiation; (2) the H· is extracted from the excited fragment RH*, resulting in the appearance of the radical molecule R· at C1 or C4 position; and (3) glycosidic bonds can be decomposed by radical molecule R·(Tissot et al., 2013).

Gamma radiation has been regarded as an effective tool to prepare and modify nano-starch. Akhavan et al. (2012) have successfully assembled starch nanoparticles under γ-irradiation with a dose of 20 kGy, and they found the size of the nanoparticle could be decreased using surfactants during the nanoparticle formation process. Lamanna et al. (2013) developed ultra-small starch nanoparticles by γ-irradiation, and cassava and waxy starches can be assembled into nanoparticles of 20 and 30 nm, respectively, which have potential application in matrix filler.

9.2.4 Milling

Milling has been employed in numerous studies to create a variety of nano-starch samples (Figure 9.2) (Chen, Shen, & Yeh, 2010; Patel, Chakraborty, & Murthy, 2016). The weight and speed of the grinding ball, which continue to move and spin during milling, are what reduce micron-sized starch granules to nano-size. Kinetic energy is created by this motion, and the sample size may decrease as a result of this action (Marta et al., 2022). Many scientists have studied how the milling process affects the micro-structure of starch. In brief, the starch granules would flake off layer-by-layer from the surface to the interior and shatter into anomalous particles when they were mechanically damaged and partially broken by the shearing and impact forces from the grinding media (Ren et al., 2010).

Although milling seems to be an economically viable technology for the production of starch nanoparticles, it creates particles of a larger size than other physical processes sometimes. Patel et al. (2016) reported employing the medium milling procedure to form nano-starch with a size of about 245 nm. Ahmad and Gani et al. (2020) prepared a starch nanoparticle with a size of 855 nm. However, they found it interesting of increased transition temperature and viscosity after the ball milling process. The shape and crystalline structure of starch molecules may be physically altered during milling, which can alter their properties, including improved cold-water dispersibility, decreased crystallinity, increased viscosity, and increased transition temperature (Lu et al., 2018).

9.3 FORMATION OF POLYMORPHIC NANO-STARCH

Nano-starches are usually tailored into different shapes with the help of physical processes to act on starch hierarchical structure, and the formed polymorphic samples including starch nano-sphere (SNS), starch nano-rod (SNR), starch nano-worm (SNW), starch nano-polyhedron (SNPH), starch nano-flake (SNF), starch nano-vesicle (SNV), starch nano-micelle (SNM), and starch nano-networks (SNNWs), etc. (Figure 9.3) (Yao et al., 2024).

9.3.1 Starch Nano-sphere

Nano-starch (often like nano-sphere) is widely prepared by nano-precipitation induced by antisolvent such as ethanol (Qin & Liu et al., 2016; Qiu & Yang et al., 2016). The generation of spherical nano-starch is mainly from the driving forces (Figure 9.4a), including surface tension changes,

FIGURE 9.3 Polymorphic nano-starch including starch nano-sphere (SNS), starch nano-rod (SNR), starch nano-worm (SNW), starch nano-polyhedron (SNPH), starch nano-flake (SNF), starch nano-vesicle (SNV), starch nano-micelle (SNM), and starch nano-networks (SNNWs).

flow, and diffusion between solute, solvent, and antisolvent molecules (Joye & McClements, 2013; Lin et al., 2022; Nicolas et al., 2013). However, the starch nano-spheres are liable to aggregate, which hinders their applications by nano-size effect (Ruan et al., 2022). Herein, chemical surfactants were generally added to modify nano-starch with enhanced dispersity. For example, Mahmoudi Najafi et al. (2016) modified corn starch with acetic anhydride and acetic acid prior to nano-precipitation and found that chemical surfactants promoted to form SNS with high dispersibility. Some physical treatments also assist the nano-precipitation method to form SNS. Yao et al. (2023) synthesized SNSs by combining nano-precipitation with DHPM. The DHPM treatment was used to control the basic structure of starch chains and then the high-uniformity-high-dispersity SNSs were separated and purified by nano-precipitation.

9.3.2 Starch Nano-rod, Starch Nano-worm, and Starch Nano-network

SNR, SNW, and SNNW are usually assembled by enzymatic hydrolysis (Figure 9.4b). Enzymatic hydrolysis adjusts the morphology of nano-starch through specific action sites on the glycosidic bonds of starch chains. A-Amylase, β-amylase, and pullulanase are widely used to break starch chains for obtaining different morphologies of nanoparticles such as SNR, SNS, and even SNW (Campelo et al., 2020; Kumari, Yadav, & Yadav, 2020; Qiu et al., 2019; Qiu et al., 2020). Among them, pullulanase is the most used and it selectively hydrolyses α-D-(1,6) glycosidic bonds to form SNR.

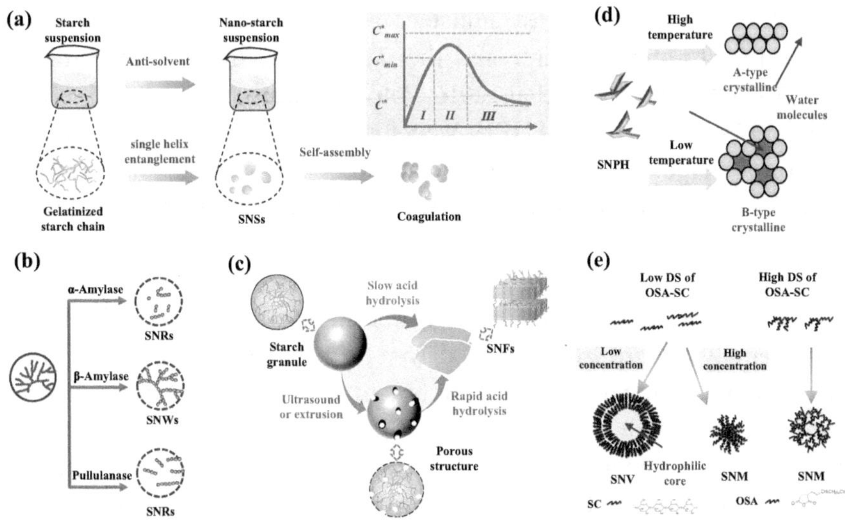

FIGURE 9.4 Formation of polymorphic nano-starch: (a) starch nano-sphere (SNS), (b) starch nano-rod (SNR) and starch nano-worm (SNW), (c) starch nano-flaket (SNF), (d) starch nano-polyhedron (SNPH), (e) starch nano-vesicle (SNV), and starch nano-micelle (SNM). Reproduction with the permission from reference Yao et al. (2024). Copyright 2024 Elsevier.

It causes an increase of crystalline regions and double helix structures (Tang et al., 2022; Wang & Lin et al., 2021; Wang & Chang et al., 2021). Besides, some mechanical methods are usually used to accelerate the reaction degree of enzyme treatment (Rostamabadi, Falsafi, & Jafari, 2019).

9.3.3 Starch Nano-polyhedron and Starch Nano-flake

Nano-starch like starch nano-polyhedron and starch nano-flake with high crystallinity could be prepared by acid hydrolysis (Figure 9.4c and d). The process of acid-induced starch hydrolysis can be summarized into three steps. (1) Fast degradation. The surface of the starch corroded and amorphous region was hydrolysed rapidly. (2) Slow degradation. The starch wall collapsed and the amorphous degraded. (3) Extremely slow degradation. Nanoparticle of the crystalline layers is gradually formed (Campelo et al., 2020; Jenkins & Donald, 2006; LeCorre, Bras, & Dufresne, 2011; Wang, Truong, & Wang, 2003). The last two stages are the key to preparing small particle sizes of SNPHs and SNFs. Many physical pre-treatments such as extrusion (Xu & Campanella et al., 2020), ultrasound, plasma (Shen et al., 2022), etc. could be utilized to open the intact structure of starch granule for strong acid action. SNPHs could also be produced by self-assembly, with different morphologies adjusted by conditions like temperature, the degree of polymerization, and the ratio of starch to solvent (Buleon et al., 1984; Kiatponglarp et al., 2016; Potocki-Veronese et al., 2005; Wu & Sarko, 1978). Interestingly, SNPHs were formed at B-type or A-type crystal structures, respectively, at low temperatures (<25°C) or high temperatures (>50°C) (Buléon, Véronèse, & Putaux, 2007; Kiatponglarp et al., 2015).

9.3.4 Starch Nano-vesicle and Starch Nano-micelle

Starch nano-vesicle and starch nano-micelle are specially produced by self-assembly (Figure 9.4e), mainly driven by thermodynamics and chemical bonds, including hydrogen bonds and intermolecular forces (Li, Fan, & Yin, 2021; Luo, Adra, & Kim, 2020). The formation of SNV and SNM (especially modified with amphiphilic fragments) go through three stages, including nucleation of dextran chains, nucleus growth, and aggregation of nanoparticles (Li et al., 2016). The DP of dextran chains depend on the formation of SNV and SNM. Chang and Yang et al. (2017) prepared short starch chains (OSA-SC) modified with octenyl succinic anhydride and revealed the correlation between degrees of substitution and the morphology of starch nanoparticle. Though almost no physical treatment has been reported to create SNV or SNM, it has potential to promote the generation of them with appropriate DP values under diverse mechanical actions.

9.4 NANO-STARCH IN FOOD AND NON-FOOD APPLICATIONS

9.4.1 Emulsion

Nano-starch is widely used as a stabilizer for emulsion (i.e., Pickering emulsion) due to its biodegradability, biocompatibility, small size, and low cost. Pickering emulsions are defined as ones stabilized by solid particles, which is opposed to ordinary emulsions containing surfactants. It can be explained in three steps (Figure 9.5) (Marefati & Rayner, 2020; Pickering & Umfreville, 1907). (1) First, the capillary force created by the interaction of nearby particles prevents the mass transfer at the contact due to interfacial pressure (Matos et al., 2017). (2) The stability of emulsions is then increased as a result of the formation of aggregates in the three-dimensional state of the emulsion droplet network, which has stemmed from the particle interaction that builds bridges between droplets (Lam, Velikov, & Velev, 2014). (3) In the continuous phase, too many particles form aggregated networks that thicken the emulsion, create a gel structure, and slow down gravity separation (Lam et al., 2014). The stability mechanism of

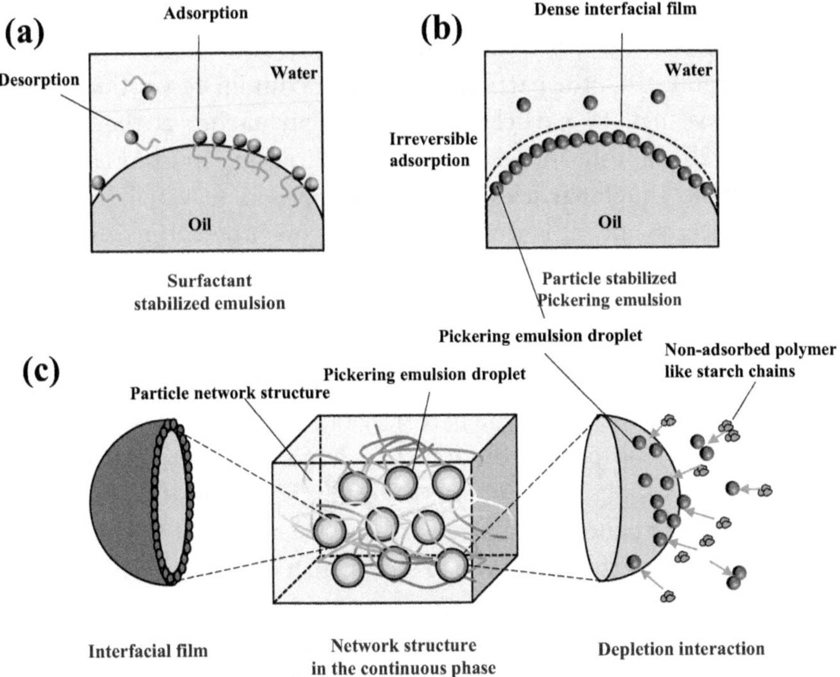

FIGURE 9.5 Stabilization mechanism of Pickering emulsion: (a) Surfactant stabilized emulsion and (b) particle-stabilized Pickering emulsion. (c) Three potential stabilization mechanisms of Pickering emulsion by starch nanoparticles.

nano-starch-based Pickering emulsion is the directional adsorption of particles at the oil–water interface and the formation of a physical energy barrier. Thus, the adsorption behaviour of nano-starch particles at the oil–water interface will directly affect the stability and functional characteristics of the emulsion. High stability is necessary for its long shelf life.

Emulsions are vulnerable to the outside environment and have a great propensity to stratify due to thermodynamic instability. Pickering emulsion destabilization has previously been proven to be closely connected to their interfacial characteristics though occasionally indirectly (Berton-Carabin, Sagis, & Schroën, 2018). The wettability and size of starch nanoparticles have a significant impact on adsorption stability, besides pH, ionic strength, particle properties, and concentrations also affect the stability of Pickering emulsion (Yao et al., 2024). The wettability of solid particles like nano-starch affects their ability to adsorb and stabilize emulsion at the interface. The adsorption of partially wetted particles leads to their aggregation at the oil–water interface, and the stability of this accumulation is closely related to the desorption/separation energy (Xu & Campanella et al., 2020). When the wetting contact angle of the three phases θ is about 90°, the particles have the maximum desorption energy, which means that the particles will irreversibly anchor at the interface (Low et al., 2020). If the interface stability of Pickering emulsion is significant, the nano-starch particles should have appropriate wettability on the interface theoretically. It is worth noting that the inherent hydrophilicity of starch particles makes them preferentially wetted by water and tend to form oil-in-water Pickering emulsion. Various starch particles of botanical origins have various surface compositions, which has a big impact on how wettable they are at the interface. The surface characteristics of native starch particles made of waxy maize, rice, and wheat were described by Li et al. (2013). Except for potato starch, which has a contact angle between 29° and 63°, hydrophobicity increases the ability to emulsify. For tapioca, corn, and sweet potato starch nanoparticles, Ge et al. (2017) discovered a virtually neutral wettability (close to 90°), which effectively prevents their desorption at the oil–water interface.

The size of nano-starch essentially affects the stability of the Pickering emulsion as well as the size of the emulsion droplet. Li et al. (2013) observed that emulsion would not be stabilized when large potato starch particles were greatly used as Pickering stabilizer. This result gave rise to a theory about Pickering stabilization, that is about smaller particles which produce more stable emulsions because they have faster adsorption kinetics and more efficient packing at the interface (Li et al., 2013; Wu & Ma, 2016). However, beyond a threshold of tiny particle size, the stability of emulsion declines because the Brownian effect is now powerful enough to influence how particles are divided at the liquid–liquid interface (Low et al., 2020; Tambe & Sharma, 1994). According to the Eq. (9.1) of Binks and Lumsdon (2001), the diameter of the emulsion droplet is relevant to the size of the starch nanoparticle, where r_e and r_p are the diameters of the emulsion droplet and starch nanoparticle size, respectively. Φ_d and ϕ_p are the volume fraction of dispersed phase and solid particle, respectively. For constant volume proportions of the dispersed phase and Pickering particles, there is a linear relationship between the emulsion drop size and the particle size, and the emulsion droplets should grow according to starch nanoparticle size. The general rule of thumb is that the particle size for Pickering stabilization should be at least one order of magnitude less than the needed droplet size to form a stable emulsion. However, Binks and Lumsdon (2001) found that this relationship eventually falters, which is most likely brought on by a change in particle contact angle, directly affecting the number of particles at the liquid–liquid interface.

$$R_e = \frac{4\phi_d r_p}{\phi_p} \tag{9.1}$$

Due to its hydrophilicity, nano-starch is typically modified to gain amphiphilic characteristics for simultaneous adsorption on different interfaces. The amphiphilicity of nano-starch is improved in the case of an oil-in-water system to create a particle layer on the surface of oil droplets. This layer prevents contact between adjacent droplets through steric hindrance and suspends oil droplets in the water (Aveyard, Binks, & Clint, 2003; Binks, 2002; Bortnowska, 2012; Dickinson, 2010; Li et al., 2018). The Pickering emulsion stabilized by taro SNSs was successfully made by Ping et al. (2018), and they methodically investigated the impact of particle concentration, oil–water ratio, and sodium chloride concentration on the stability of emulsions. The stability of the emulsion was the greatest when 7% particle concentration and 0.5 oil fraction were used in the above study. Tea polyphenols were encapsulated with a 67% loading rate after slight flocculation at a low salt ion concentration (0.04 mM NaCl)., Lee et al. (2021) encapsulated curcumin in a Pickering emulsion stabilized by SNSs and chitin nano-fibres. Curcumin retention considerably increased (after 9 hours, from 65% to 80%) when the solid concentration rose (from 1% SNSs and 0.2% chitin nano-fibres to 2% SNSs and 0.4% chitin nano-fibres). Some physical treatments also help stabilize Pickering emulsion. Mechanical forces such as homogenization and ultrasonication were utilized for the starch nanoparticle to stabilize the emulsion (Ruan et al., 2022). Time prolongation of homogenization led to a smaller emulsion droplet, while ultrasonication with periodic cavitation could disintegrate starch nanoparticle into micro-aggregates for a stable emulsion system in a short period.

9.4.2 Delivery System

Compared with direct intake of bioactive substances, or those assisted by traditional inorganic carriers, we may achieve a higher and safer bioavailability of bioactives based on a nano-starch delivery system for targeted delivery and controlled release (Yuan et al., 2021). So far, nano-starch-mediated delivery systems are usually used to simulate oral release in humans (Figure 9.6), with the oral cavity, stomach, small intestine, and colon

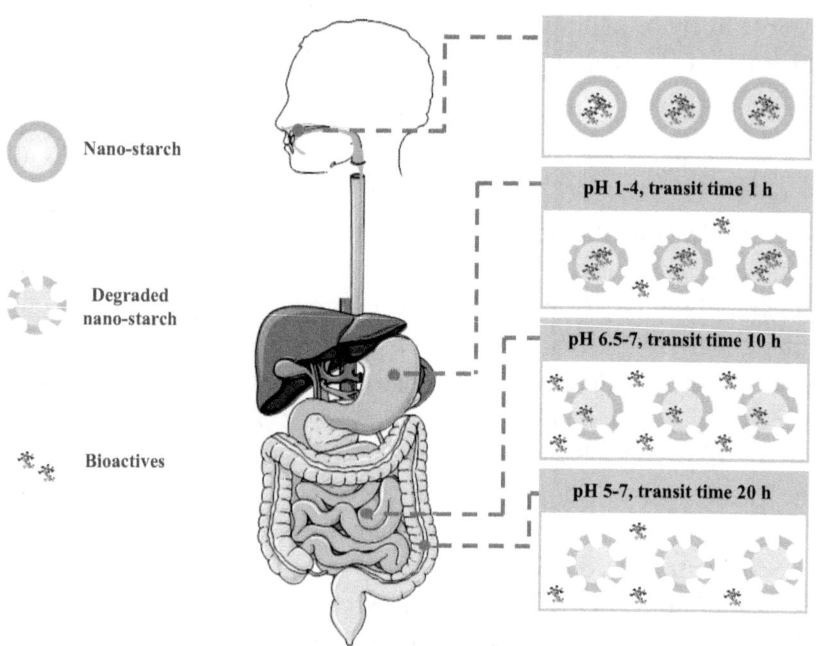

FIGURE 9.6 Simulated in-vivo release of bioactive compound from nano-starch in the oral cavity, stomach, small intestine, and colon.

at pH 5–7, 1–4, 6.5–7, and 5–7, respectively (Matilde, Niu, & Alonso, 2020). The compact structure of complexes assembled by intermolecular force may to some extent protect the bioactive compounds from degradation and pre-mature release in gastric microenvironment. The progress of the targeted delivery of different bioactives via nano-starch and their controlled release features are mainly introduced in this section.

Biopolymers are easily bound to target sites due to the presence of (or modified) surface functional groups for targeted delivery (Verma et al., 2020). Nano-starch is rich in hydroxyl groups exposed on the surface to bind with targets as well as other modified functional groups such as carboxyl, carbonyl, and ester carbonyl groups (Alberto et al., 2020; Leonardo et al., 2018; Miskeen et al., 2021). Nano-starch and other carriers could enter the digestive system and then the blood circulation through biological barriers to reach targeted sites, which is accomplished by active targeting, passive targeting, or stimulus-regulated release (Figure 9.7) (Chamundeeswari, Jeslin, & Verma, 2019). In active targeting, ligands are grafted onto nano-carriers, and they penetrate healthy tissues through intracellular transfer, transmembrane channels, or paracellular transport

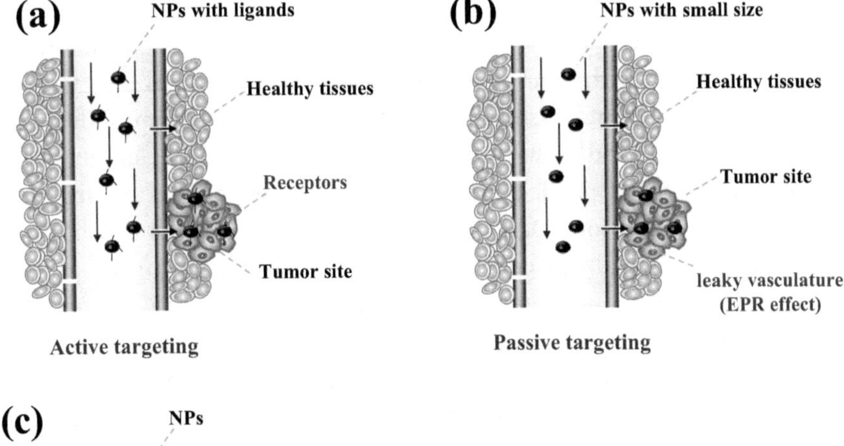

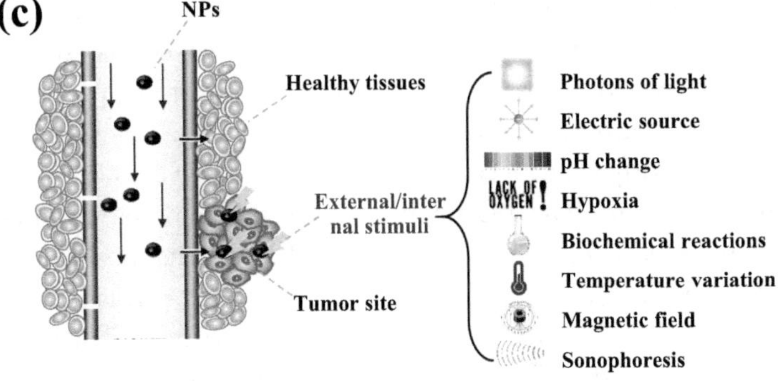

FIGURE 9.7 Potential targeted release based on starch nanoparticles (NS) through (a) active targeting, (b) passive targeting, or (c) stimulus-regulated release.

systems, which promote the recognition of ligands-loaded nano-carriers by cell surface receptors (Figure 9.7a) (Wang & Thanou, 2010; Yoo et al., 2019). The ligands include folic acid, polysaccharides, lipoproteins, peptides, etc. (Danhier, Feron, & Préat, 2010; Pérez-Herrero & Fernández-Medarde, 2015). The success of active targeting mainly depends on the type of targeted cells (Jain et al., 2013; Zhan et al., 2012). In passive targeting (Figure 9.7b), nano-carriers pass through blood vessels with pore sizes ranging from 100 nm to 2 μm and accumulate on target sites (Gaumet et al., 2008). Passive targeting has strict requirements for nano-carriers. To avoid being filtered by the kidney and captured by the liver, nano-carriers need to be controlled between about 10 and 100 nm. The average particle size of self-assembled nano-starch may achieve 100 nm or even smaller than 50 nm, which has the potential to transport bioactives by passive targeting. Moreover, mechanical force is conducive to the acquisition of nano-starch with a particle size of less than 50 nm (Ruan et al., 2022). Compared with macroparticle carriers, nanoparticles are easier to passively aggregate into tissue cells due to the enhanced permeability and retention effect (so-called EPR effect) (Danhier et al., 2010; Gaumet et al., 2008). The charge of particles should be neutral or anionic to effectively avoid renal elimination (Danhier et al., 2010). Nano-starch can be constructed with neutral or negative surfaces by modification such as succinylation, neutral, and anionic surfactant. Besides, bioactive molecules encapsulated in nano-starch can be released through internal or external stimuli such as photons of light, electric source, pH, hypoxia, biochemical reactions, temperature variation, magnetic field, sonophoresis, etc. (Figure 9.7c) (Verma et al., 2020). The stimulation response of nano-starch and its modified derivatives is particularly critical. However, the precisely targeted delivery by nano-starch is still in the initial stage, considering that bioactives are mostly influenced by pH or enzymes (see below).

Two chemical strategies have mainly been used for the preparation of pH-sensitive delivery systems. One is to utilize pH-labile chemical bonds such as hydrazone, maleic, and acetal bonds (Fleige, Quadir, & Haag, 2012; Shi et al., 2020; Xu & Zi et al., 2020), and another is to use "titratable" groups such as amine, carboxyl, and imidazole functional groups (Fleige et al., 2012; Ganta et al., 2008). Both of them induce the physical dissociation or internal structure change of the complex through the change of pH value, so as to achieve the targeted release of active compounds. The latter strategy is more common in nano-starch carrier-mediated delivery systems. For example, Thomas et al. (2021) prepared alginate-modified nano-starch that encapsulated theophylline and bovine serum albumin, separately and released them into human bodies with the pH response ability of carboxyl groups.

In the gastric position with a low pH value, the carboxyl groups were protonated and the network structure was maintained by the strong hydrogen bonds, which protected complexes from an acidic microenvironment. When moving to the gut with an elevated pH value, the carboxyl groups were partially ionized, the nano-spheres degraded, and the objects started to release. Over time, they move to the colon where the pH is higher, and nano-starch can be mostly degraded (alginic acid is converted to its sodium salt and dissolved in water), resulting in the complete release of these bioactive compounds (Thomas et al., 2021). The complexes may be able to avoid the influence of the stomach under the mediation of pH and target the intestinal environment for sustained release. Xu and Zi et al. (2020) constructed pH-sensitive nano-carriers of cholesterol/imidazole-modified oxidized starch (Cho-Imi-OS) with the pH-triggered ability of imidazole groups to targeted release of curcumin (Cur) into cancer cells by EPR effect. These Cur-NPs with loading efficiency (4.16%) released curcumin faster at pH 5.5 than that at pH 7.4 after endocytosis by endosomes.

There are a variety of digestive enzymes in the human *in-vivo* micro-environment, which may promote or inhibit the structural degradation of complexes, and conversely it is enlightening to construct delivery systems responsive to these enzymes. Remanan and Zhu (2021) used starch nano-carriers (approximately 100–200 nm) to load rutin (>60%) and simulated the release rule of rutin *in vitro*. The inhibitory effect of rutin on salivary amylase prevented the digestion of starch nano-carriers in the mouth. In the simulated process of gastric digestion, rutin was released suddenly and continuously. However, the *in-vivo* research on nano-starch delivery system still needs to be strengthened, and whether the enzyme as a protein interferes with the binding of nano-starch and bioactive is also questionable.

REFERENCE, BIBLIOGRAPHY OR WORKS CITED

Ahmad, A. N., Lim, S. A., Navaranjan, N., Hsu, Y. I., & Uyama, H. 2020. Green sago starch nanoparticles as reinforcing material for green composites. *Polymer* 202: 122646.

Ahmad, M., Gani, A., Masoodi, F. A., & Rizvi, S. H. 2020. Influence of ball milling on the production of starch nanoparticles and its effect on structural, thermal and functional properties. *International Journal of Biological Macromolecules* 151: 85–91.

Akhavan, A., & Ataeevarjovi, E. 2012. The effect of gamma irradiation and surfactants on the size distribution of nanoparticles based on soluble starch. *Radiation Physics and Chemistry* 81 (7): 913–914.

Alberto, E. P., Adriana, G. G., Susana, R., Alejandro, Z., & Fernando, M. B. 2020. Effect of amylose/amylopectin content and succinylation on properties of corn starch nanoparticles as encapsulants of anthocyanins. *Carbohydrate Polymer* 250: 116972.

Alkanawati, M., Shafee, W., Frederik, R., Therien-Aubin, H., Landfester, K. 2018. Large-scale preparation of polymer nanocarriers by high-pressure microfluidization. *Macromolecular Materials & Engineering* 303 (1): 1700505.

Apostolidis, E., & Mandala, I. 2020. Modification of resistant starch nanoparticles using high-pressure homogenization treatment. *Food Hydrocolloids* 103: 105677.

Aveyard, R., Binks, B. P., & Clint, J. H. 2003. Emulsions stabilised solely by colloidal particles. *Advances in Colloid and Interface Science* 100–102 (28): 503–546.

Bajer, D. 2023. Nano-starch for food applications obtained by hydrolysis and ultrasonication methods. *Food Chemistry* 402: 134489.

Bel Haaj, S., Magnin, A., Pétrier, C., & Boufi, S. 2013. Starch nanoparticles formation via high power ultrasonication. *Carbohydrate Polymers* 92 (2): 1625–1632.

Bel Haaj, S., Thielemans, W., Magnin, A., & Boufi, S. 2016. Starch nanocrystals and starch nanoparticles from waxy maize as nanoreinforcement: A comparative study. *Carbohydrate Polymers* 143: 310–317.

Berton-Carabin, C. C., Sagis, L., & Schroën, K. 2018. Formation, structure, and functionality of interfacial layers in food emulsions. *Annual Review of Food Science and Technology* 9 (1): 551–587.

Binks, B. P. 2002. Particles as surfactants-similarities and differences. *Current Opinion in Colloid & Interface Science* 7 (1): 21–41.

Binks, B. P., & Lumsdon, S. O. 2001. Pickering emulsions stabilized by monodisperse latex particles: Effects of particle size. *Langmuir* 17 (15): 4540–4547.

Bitik, A., Sumnu, G., & Oztop, M. 2019. Physicochemical and structural characterization of microfluidized and sonicated legume starches. *Food and Bioprocess Technology* 12 (7): 1144–1156.

Bortnowska, G. 2012. Effects of pH and ionic strength of NaCl on the stability of diacetyl and (-)-α-pinene in oil-in-water emulsions formed with food-grade emulsifiers. *Food Chemistry* 135 (3): 2021–2028.

Buleon, A., Duprat, F., Booy, F. P., & Chanzy, H. 1984. Single crystals of amylose with a low degree of polymerization. *Carbohydrate Polymers* 4 (3): 161–173.

Buléon, A., Véronèse, G., & Putaux, J. L. 2007. Self-association and crystallization of amylose. *Australian Journal of Chemistry* 60 (10): 706–718.

Campelo, P. H., Sant'Ana, A. S., & Pedrosa Silva Clerici, M. T. 2020. Starch nanoparticles: Production methods, structure, and properties for food applications. *Current Opinion in Food Science* 33: 136–140.

Chamundeeswari, M., Jeslin, J., & Verma, M. L. 2019. Nanocarriers for drug delivery applications. *Environmental Chemistry Letters* 17 (2): 849–865.

Chang, R., Lu, H., Tian, Y., Li, H., Wang, J., & Jin, Z. 2020. Structural modification and functional improvement of starch nanoparticles using vacuum cold plasma. *International Journal of Biological Macromolecules* 145: 197–206.

Chang, R., Yang, J., Ge, S., Zhao, M., Liang, C., Xiong, L., & Sun, Q. 2017. Synthesis and self-assembly of octenyl succinic anhydride modified short glucan chains based amphiphilic biopolymer: Micelles, ultrasmall micelles, vesicles, and lutein encapsulation/release. *Food Hydrocolloids* 67: 14–26.

Chang, Y., Yan, X., Wang, Q., Ren, L., Tong, J., & Zhou, J. 2017. High efficiency and low cost preparation of size controlled starch nanoparticles through ultrasonic treatment and precipitation. *Food Chemistry* 227: 369–375.

Chang, Y., Yang, J., Jiang, L., Ren, L., & Zhou, J. 2019. Chain length distribution of β-amylase treated potato starch and its effect on properties of starch nanoparticles obtained by nanoprecipitation. *Starch-Starke* 71 (9–10): 1800321.

Chen, C. J., Shen, Y. C., & Yeh, A. I. 2010. Physico-chemical characteristics of media-milled corn starch. *Journal of Agricultural and Food Chemistry* 58 (16): 9083–9091.

Chen, L., Dai, Y., Hou, H., Wang, W., Ding, X., Zhang, H., Li, X., & Dong, H. 2021. Effect of high pressure microfluidization on the morphology, structure and rheology of sweet potato starch. *Food Hydrocolloids* 115: 106606.

Chu, D., Dong, X., Shi, X., Zhang, C., & Wang, Z. 2018. Neutrophil-based drug delivery systems. *Advanced Materials* 30 (22): 1706245.

Danhier, F., Feron, O., & Préat, V. 2010. To exploit the tumor microenvironment: Passive and active tumor targeting of nanocarriers for anti-cancer drug delivery. *Journal of Controlled Release* 148 (2):135–146.

Dickinson, E. 2010. Food emulsions and foams: Stabilization by particles. *Current Opinion in Colloid & Interface Science* 15 (1): 40–49.

Dong, H., Zhang, Q., Gao, J., Chen, L., & Vasanthan, T. 2021. Preparation and characterization of nanoparticles from cereal and pulse starches by ultrasonic-assisted dissolution and rapid nanoprecipitation. *Food Hydrocolloids* 122: 107081.

Fleige, E., Quadir, M. A., & Haag, R. 2012. Stimuli-responsive polymeric nanocarriers for the controlled transport of active compounds: Concepts and applications. *Advanced Drug Delivery Reviews* 64 (9): 866–884.

Ganta, S., Devalapally, H., Shahiwala, A., & Amiji, M. 2008. A review of stimuli-responsive nanocarriers for drug and gene delivery. *Journal of Controlled Release* 126 (3): 187–204.

Gaumet, M., Vargas, A., Gurny, R., & Delie, F. 2008. Nanoparticles for drug delivery: The need for precision in reporting particle size parameters. *European Journal of Pharmaceutics and Biopharmaceutics* 69 (1): 1–9.

Ge, S., Xiong, L., Li, M., Liu, J., Yang, J., Chang, R., Liang, C., & Sun, Q. 2017. Characterizations of Pickering emulsions stabilized by starch nanoparticles: Influence of starch variety and particle size. *Food Chemistry* 234: 339–347.

Guo, Z., Zeng, S., Lu, X., Zhou, M., Zheng, M., & Zheng, B. 2015. Structural and physicochemical properties of lotus seed starch treated with ultra-high pressure. *Food Chemistry* 186: 223–230.

Hao, Y., Yun, C., Qian, L., & Gao, Q. 2018. Preparation of starch nanocrystals through enzymatic pretreatment from waxy potato starch. *Carbohydrate Polymers* 184: 171–177.

He, X. H., Luo, S. J., Chen, M. S., Xia, W., Chen, J., & Liu, C. M. 2020. Effect of industry-scale microfluidization on structural and physicochemical properties of potato starch. *Innovative Food Science & Emerging Technologies* 60: 102278.

Jain, A., Singhai, P., Gurnany, E., Updhayay, S., & Mody, N. 2013. Transferrin-tailored solid lipid nanoparticles as vectors for site-specific delivery of temozolomide to brain. *Journal of Nanoparticle Research* 15 (3): 399–403.

Jenkins, P. J., & Donald, A. M. 2006. The effect of acid hydrolyis on native starch granule structure. *Starch-Starke* 49 (7–8): 262–267.

Joye, I. J., & McClements, D. J. 2013. Production of nanoparticles by anti-solvent precipitation for use in food systems. *Trends in Food Science & Technology* 34 (2): 109–123.

Kasemwong, K., Ruktanonchai, U. R., Srinuanchai, W., Itthisoponkul, T., & Sriroth, K. 2011. Effect of high-pressure microfluidization on the structure of cassava starch granule. *Starch-Starke* 63 (3): 160–170.

Kiatponglarp, W., Rugmai, S., Rolland-Sabaté, A., Buléon, A., & Tongta, S. 2016. Spherulitic self-assembly of debranched starch from aqueous solution and its effect on enzyme digestibility. *Food Hydrocolloids* 55: 235–243.

Kiatponglarp, W., Tongta, S., Rolland-Sabaté, A., & Buléon, A. 2015. Crystallization and chain reorganization of debranched rice starches in relation to resistant starch formation. *Carbohydrate Polymers* 122: 108–114.

Kim, H. Y., Park, D. J., Kim, J. Y., & Lim, S. T. 2013. Preparation of crystalline starch nanoparticles using cold acid hydrolysis and ultrasonication. *Carbohydrate Polymers* 98 (1): 295–301.

Koubaa, M., Barba-Orellana, S., Roselló-Soto, E., & Barba, F. J. 2016. Gamma irradiation and fermentation. Novel Food Fermentation Technologies. K. S. Ojha, and B. K. Tiwari, eds., 143–153. Cham: Springer International Publishing.

Kumari, S., Yadav, B. S., & Yadav, R. B. 2020. Synthesis and modification approaches for starch nanoparticles for their emerging food industrial applications: A review. *Food Research International* 128: 108765.

Lam, S., Velikov, K. P., & Velev, O. D. 2014. Pickering stabilization of foams and emulsions with particles of biological origin. *Current Opinion in Colloid & Interface Science* 19 (5): 490–500.

Lamanna, M., Morales, N. J., García, N. L., & Goyanes, S. 2013. Development and characterization of starch nanoparticles by gamma radiation: Potential application as starch matrix filler. *Carbohydrate Polymers* 97 (1): 90–97.

LeCorre, D., Bras, J., & Dufresne, A. 2011. Evidence of micro- and nanoscaled particles during starch nanocrystals preparation and their isolation. *Biomacromolecules* 12 (8): 3039–3046.

Lee, Y. S., Tarte, R., & Acevedo, N. C. 2021. Curcumin encapsulation in Pickering emulsions co-stabilized by starch nanoparticles and chitin nanofibers. *RSC Advances* 11: 16275–16284.

Leonardo, A. G., Leonardo, N. S., Leidy, T. S., Pinzon, M. I., & Villa, C. C. 2018. Development of native and modified banana starch nanoparticles as vehicles for curcumin. *International Journal of Biological Macromolecules* 111: 498–504.

Li, C., Li, Y., Sun, P., & Yang, C. 2013. Pickering emulsions stabilized by native starch granules. *Colloids and Surfaces A: Physicochemical and Engineering Aspects* 431: 142–149.

Li, J., Ye, F., Lei, L., Zhou, Y., & Zhao, G. 2018. Joint effects of granule size and degree of substitution on octenylsuccinated sweet potato starch granules as pickering emulsion stabilizers. *Journal of Agricultural and Food Chemistry* 66 (17): 4541–4550.

Li, X., Qin, Y., Liu, C., Jiang, S., Xiong, L., & Sun, Q. 2016. Size-controlled starch nanoparticles prepared by self-assembly with different green surfactant: The effect of electrostatic repulsion or steric hindrance. *Food Chemistry* 199: 356–363.

Li, Z., Fan, Q., & Yin, Y. 2021. Colloidal self-assembly approaches to smart nanostructured materials. *Chemical Reviews* 122 (5): 4976–5067.

Lin, Q., Liu, Y., Zhou, L., Ji, N., Xiong, L., & Sun, Q. 2022. Green preparation of debranched starch nanoparticles with different crystalline structures by electrostatic spraying. *Food Hydrocolloids* 127: 107513.

Low, L. E., Siva, S. P., Ho, Y. K., Chan, E. S., & Tey, B. T. 2020. Recent advances of characterization techniques for the formation, physical properties and stability of pickering emulsion. *Advances in Colloid and Interface Science* 277: 102117.

Lu, X., Xiao, J., & Huang, Q. 2018. Pickering emulsions stabilized by media-milled starch particles. *Food Research International* 105: 140–149.

Luo, K., Adra, H. J., & Kim, Y. R. 2020. Preparation of starch-based drug delivery system through the self-assembly of short chain glucans and control of its release property. *Carbohydrate Polymers* 243: 116385.

Mahmoudi Najafi, S. H., Baghaie, M., & Ashori, A. 2016. Preparation and characterization of acetylated starch nanoparticles as drug carrier: Ciprofloxacin as a model. *International Journal of Biological Macromolecules* 87: 48–54.

Marefati, A., & Rayner, M. 2020. Starch granule stabilized pickering emulsions: An 8-year stability study. *Journal of the Science of Food and Agriculture* 100 (6): 2807–2811.

Marta, H., Wijaya, C., Sukri, N., Cahyana, Y., & Mohammad, M. 2022. A comprehensive study on starch nanoparticle potential as a reinforcing material in bioplastic. *Polymers* 14 (22): 4875.

Matilde, D. L., Niu, Z., & Alonso, M. J. 2020. Oral delivery of biologics for precision medicine. *Advanced Materials* 32 (13): 1901935.

Matos, M., Marefati, A., Bordes, R., Gutiérrez, G., & Rayner, M. 2017. Combined emulsifying capacity of polysaccharide particles of different size and shape. *Carbohydrate Polymers* 169: 127–138.

Mei, J. Y., Zhang, L., Ren, M. H., Lin, Y., & Fu, Z. 2020. Insight into multi-scale structure and digestibility of sugar palm (*Arenga pinnata*) starch subjected to high speed jet treatment. Starch-Stärke 72 (9–10): 1900278.

Miskeen, S., An, Y. S., & Kim, J. Y. 2021. Application of starch nanoparticles as host materials for encapsulation of curcumin: Effect of citric acid modification. *International Journal of Biological Macromolecules* 183: 1–11.

Nicolas, J., Mura, S., Brambilla, D., Mackiewicz, N., & Couvreur, P. 2013. Design, functionalization strategies and biomedical applications of targeted biodegradable/biocompatible polymer-based nanocarriers for drug delivery. *Chemical Society Reviews* 42 (3): 1147–1235.

Ozturk, O. K., & Turasan, H. 2021. Applications of microfluidization in emulsion-based systems, nanoparticle formation, and beverages. *Trends in Food Science & Technology* 116: 609–625.

Patel, C. M., Chakraborty, M., & Murthy, Z. V. P. 2016. Fast and scalable preparation of starch nanoparticles by stirred media milling. *Advanced Powder Technology* 27 (4): 1287–1294.

Pérez-Herrero, E., & Fernández-Medarde, A. 2015. Advanced targeted therapies in cancer: Drug nanocarriers, the future of chemotherapy. *European Journal of Pharmaceutics and Biopharmaceutics* 93: 52–79.

Pickering, S. U. 1907. Emulsions. *Journal of the Chemical Society Transaction* 91: 2001–2021.

Ping, S., Zhang, H., Niu, B., & Jin, W. 2018. Physical stabilities of taro starch nanoparticles stabilized pickering emulsions and the potential application of encapsulated tea polyphenols. *International Journal of Biological Macromolecules* 118 (15): 2032–2039.

Potocki-Veronese, G., Putaux, J. L., Dupeyre, D., Albenne, C., Remaud-Siméon, M., Monsan, P., & Buleon, A. 2005. Amylose synthesized in vitro by amylosucrase: Morphology, structure, and properties. *Biomacromolecules* 6 (2): 1000–1011.

Qiu, C., Hu, Y., Jin, Z., McClements, D. J., Qin, Y., Xu, X., & Wang, J. 2019. A review of green techniques for the synthesis of size-controlled starch-based nanoparticles and their applications as nanodelivery systems. *Trends in Food Science & Technology* 92: 138–151.

Qin, Y., Liu, C., Jiang, S., Xiong, L., & Sun, Q. 2016. Characterization of starch nanoparticles prepared by nanoprecipitation: Influence of amylose content and starch type. *Industrial Crops and Products* 87: 182–190.

Qiu, C., Wang, C., Gong, C., McClements, D. J., Jin, Z., & Wang, J. 2020. Advances in research on preparation, characterization, interaction with proteins, digestion and delivery systems of starch-based nanoparticles. *International Journal of Biological Macromolecules* 152: 117–125.

Qiu, C., Yang, J., Ge, S., Chang, R., Xiong, L., & Sun, Q. 2016. Preparation and characterization of size-controlled starch nanoparticles based on short linear chains from debranched waxy corn starch. *LWT* 74: 303–310.

Remanan, M. K., & Zhu, F. 2021. Encapsulation of rutin using quinoa and maize starch nanoparticles. *Food Chemistry* 353: 128534.

Ren, G. Y., Li, D., Wang, L. J., Özkan, N., & Mao, Z. H. 2010. Morphological properties and thermoanalysis of micronized cassava starch. *Carbohydrate Polymers* 79 (1): 101–105.

Rostamabadi, H., Falsafi, S. R., & Jafari, S. M. 2019. Starch-based nanocarriers as cutting-edge natural cargos for nutraceutical delivery. *Trends in Food Science & Technology* 88: 397–415.

Ruan, S., Tang, J., Qin, Y., Wang, J., Yan, T., Zhou, J., Gao, D., Xu, E., & Liu, D. 2022. Mechanical force-induced dispersion of starch nanoparticles and nanoemulsion: Size control, dispersion behaviour, and emulsified stability. *Carbohydrate Polymers* 275: 118711.

Shen, H., Ge, X., Zhang, B., Su, C., Zhang, Q., Jiang, H., Zhang, G., Yuan, L., Yu, X., & Li, W. 2022. Preparing potato starch nanocrystals assisted by dielectric barrier discharge plasma and its multiscale structure, physicochemical and rheological properties. *Food Chemistry* 372 :131240.

Shi, Z., Zhou, Y., Fan, T., Lin, Y., Zhang, H., & Mei, L. 2020. Inorganic nano-carriers based smart drug delivery systems for tumor therapy. *Smart Materials in Medicine* 1: 32–47.

Singh, S., Singh, N., Ezekiel, R., & Kaur, A. 2011. Effects of gamma-irradiation on the morphological, structural, thermal and rheological properties of potato starches. *Carbohydrate Polymers* 83 (4): 1521–1528.

Tambe, D. E., & Sharma, M. M. 1994. The effect of colloidal particles on fluid-fluid interfacial properties and emulsion stability. *Advances in Colloid and Interface Science* 52: 1–63.

Tang, J., Zhou, J., Zhou, X., Li, D., Wu, Z., Tian, J., Xu, E., & Liu, D. 2022. Rearranged supramolecular structure of resistant starch with polymorphic microcrystals prepared in high-solid enzymatic system. *Food Hydrocolloids* 124: 107215.

Thomas, D., Mathew, N., & Nath, M. S. 2021. Starch modified alginate nanoparticles for drug delivery application. *International Journal of Biological Macromolecules* 173: 277–284.

Tissot, C., Grdanovska, S., Barkatt, A., Silverman, J., & Al-Sheikhly, M. 2013. On the mechanisms of the radiation-induced degradation of cellulosic substances. *Radiation Physics and Chemistry* 84: 185–190.

Verma, M. L., Dhanya, B. S., Sukriti, Rani, V., Thakur, M., Jeslin, J., & Kushwaha, R. 2020. Carbohydrate and protein based biopolymeric nanoparticles: Current status and biotechnological applications. *International Journal of Biological Macromolecules* 154: 390–412.

Wang, B., Lin, X., Zheng, Y., Zeng, M., Huang, M., & Guo, Z. 2021. Effect of homogenization-pressure-assisted enzymatic hydrolysis on the structural and physicochemical properties of lotus-seed starch nanoparticles. *International Journal of Biological Macromolecules* 167: 1579–1586.

Wang, F., Chang, R., Ma, R., Qiu, H., & Tian, Y. 2021. Eco-friendly and pH-responsive nano-starch-based superhydrophobic coatings for liquid-food residue reduction and freshness monitoring. *ACS Sustainable Chemistry & Engineering* 9 (30): 10142–10153.

Wang, M., & Thanou, M. 2010. Targeting nanoparticles to cancer. *Pharmacological Research* 62 (2): 90–99.

Wang, R., Li, S., Lin, H., Yang, E., Song, Z., Zhu, B., Qin, L., & Qiu, Y. 2023. Experimental study on the influence of ultrasonic excitation duration on the pore and fracture structure and permeability of coal body. *Geoenergy Science and Engineering* 231: 212393.

Wang, Y. J., Truong, V. D., & Wang, L. 2003. Structures and rheological properties of corn starch as affected by acid hydrolysis. *Carbohydrate Polymers* 52 (3): 327–333.

Wu, H. C. H., & Sarko, A. 1978. The double-helical molecular structure of crystalline a-amylose. *Carbohydrate Research* 61 (1): 27–40.

Wu, J., & Ma, G. H. 2016. Recent studies of pickering emulsions: Particles make the difference. *Small* 12 (34): 4633–4648.

Wu, X., Zhao, L., Fang, F., Guo, Y., Liang, W., Ma, Y., & Wang, L. 2021. Progress in understanding the interaction between bioactive components of soybean and gut microbiota. *Food Science* 42 (13): 265–272.

Xu, E., Campanella, O. H., Ye, X., Jin, Z., Liu, D., & BeMiller, J. N. 2020. Advances in conversion of natural biopolymers: A reactive extrusion (REX)-enzyme-combined strategy for starch/protein-based food processing. *Trends in Food Science & Technology* 99: 167–180.

Xu, T., Yang, J., Hua, S., Hong, Y., Gu, Z., Cheng, L., Li, Z., & Li, C. 2020. Characteristics of starch-based pickering emulsions from the interface perspective. *Trends in Food Science & Technology* 105: 334–346.

Xu, Y., Zi, Y., Lei, J., Mo, X., Shao, Z., Wu, Y., Tian, Y., Li, D., & Mu, C. 2020. pH-responsive nanoparticles based on cholesterol/imidazole modified oxidized-starch for targeted anticancer drug delivery. *Carbohydrate Polymers* 233: 115858.

Yao, S., Ma, S., Zhu, Q., Qin, Y., Ngah, W. Y., Zuo, Y., Liu, Y., Zhang, X., Tian, J., Kong, X., Liu, D., & Xu, E. 2023. Ultrasmooth and uniform starch nanosphere with new microstructure formation via microfluidization-nanoprecipitation control. *ACS Sustainable Chemistry & Engineering* 11 (19): 7475–7488.

Yao, S., Zhu, Q., Xianyu, Y., Liu, D., & Xu, E. 2024. Polymorphic nanostarch-mediated assembly of bioactives. *Carbohydrate Polymers* 324: 121474.

Yan, X. X., Diao, M., Yu, Y., Gao, F., Wang, E., Wang, Z., Zhang, T., & Zhao, P. 2021. Characterization of resistant starch nanoparticles prepared via debranching and nanoprecipitation. *Food Chemistry* 369 (1): 130824.

Ye, J., Luo, S., Huang, A., Chen, J., Liu, C., & McClements, D. J. 2019. Synthesis and characterization of citric acid esterified rice starch by reactive extrusion: A new method of producing resistant starch. *Food Hydrocolloids* 92: 135–142.

Yoo, J., Park, C., Yi, G., Lee, D., & Koo, H. 2019. Active targeting strategies using biological ligands for nanoparticle drug delivery systems. *Cancers* 11 (5): 640.

Yu, L., Zhao, A., Yang, M., Wang, C., Wang, M., & Bai, X. 2018. Effects of the combination of freeze-thawing and enzymatic hydrolysis on the microstructure and physicochemical properties of porous corn starch. Food Hydrocolloids 83 (8): 465–472.

Yu, Y., & Wang, J. 2007. Effect of γ-ray irradiation on starch granule structure and physicochemical properties of rice. *Food Research International* 40 (2): 297–303.

Yuan, Y., He, N., Dong, L., Guo, Q., Zhang, X., Li, B., & Li, L. 2021. Multiscale shellac-based delivery systems: From macro- to nanoscale. *ACS Nano* 15 (12): 18794–18821.

Yun, C., Hao, Y., Kou, T., Qian, L., & Gao, Q. 2019. Preparation and emulsification properties of dialdehyde starch nanoparticles. *Food Chemistry* 286: 467–474.

Zhan, C., Wei, X., Jun, Q., Feng, L., Zhu, J. Z., & Lu, W. L. A. 2012. Co-delivery of TRAIL gene enhances the anti-glioblastoma effect of paclitaxel in vitro and in vivo. *Journal of Controlled Release* 160 (3): 630–636.

Zhu, F. 2017. Encapsulation and delivery of food ingredients using starch based systems. *Food Chemistry* 229 (15): 542–552.

Zhu, F. 2021. Structure and physicochemical properties of starch affected by dynamic pressure treatments: A review. *Trends in Food Science & Technology* 116: 639–654.

Zuo, Y., Liu, W., Xiao, J., Zhao, X., Zhu, Y., & Wu, Y. 2017. Preparation and characterization of dialdehyde starch by one-step acid hydrolysis and oxidation. *International Journal of Biological Macromolecules* 103: 1257–1264.

CHAPTER 10

Bottom-Up Additive Manufacturing of Starch-Added Materials and Starch Gels

Jie Li, Huifang Cao, Enbo Xu, and Haibo Pan

10.1 INTRODUCTION

In 1986, the first 3D printer was developed, and since then, additive manufacturing, i.e. 3D printing technology has experienced tremendous changes from the initial sprout to the present stage of rapid development in extensive fields (Ngo et al. 2018). The application scenarios of 3D printing are gradually increased from industrial design, construction, automotive, to home appliances, health care, apparel, food, education, cultural relics restoration, and even aerospace (Shahrubudin, Lee, & Ramlan, 2019; Jandyal et al., 2022; Chen et al., 2022; Mantihal, Kobun, & Lee, 2020). They use a 3D construction strategy via digital printing to achieve a 3D object, with the control of the laser, inkjet or extrusion equipment via a specify path (Rong et al., 2023). Recently, starch-related materials were also used for 3D printing (Figure 10.1), especially for food applications as well as biomedicine and environment, enabling users to gain an unprecedented level of personalization, flexibility, and engagement with their creations in developing starch products.

DOI: 10.1201/9781003493594-10

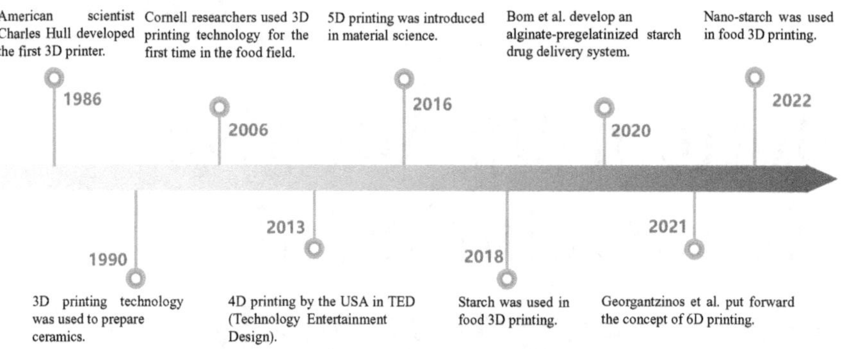

FIGURE 10.1 Timeline of important events held in the field of 3D food printing.

Thanks to the derivative invention of 3D printing technology and machine, many manufacturing advancements in food (and non-food 3D products, but less from starch materials) are realized now (Tian, Bryksa, & Yada, 2016). Due to the rising of cutting-edge 3D printing design, it is possible to create and put together foods or medicines with tailored nutrition and structure (Xu et al., 2023), as well as to shorten production and supply chains. One of the emphases has been focused on customization and diversity, fulfilling the special needs of particular customer and patient groups (Chen, Xie et al., 2019). For instance, it benefits those who have trouble swallowing (Lorenz et al., 2022) or some kids who fail to receive enough vitamins and nutrients (Derossi et al., 2018; Herrada-Manchón et al., 2020). Also, this technology of shaping 3D food with balanced ingredients may help to support the astronauts for upcoming lengthy space trips where they only have access to scarce food resources and unglamorous diet types (Enfield et al., 2022; Obrist et al., 2019; Kim & Rhee, 2020), or help them to construct temporary camp and base by 3D printing.

Starch is an abundant raw material suitable for 3D printing, either as a primary printing ink or as an additive for improving the printability of base materials. It has great potential as green and even edible printing ink. Generally, heating imparts the viscoelasticity of gelatinized starch gel (avoiding the natural impact and hierarchical structure of rigid starch particles to block 3D nozzle or induce inhomogeneous extrusion), allowing it to quickly respond to applied shear strains and thin under shear (Ji et al., 2022; Zheng et al., 2022). The retrogradation of starch chains as bottom-up components to some extent improves the mechanical strength of the whole starch-containing gel, which aids in the retention of the intended printing shape (Rong et al., 2022). In this chapter, we want to elaborate on the patterns and principles of 3D printing as a revolutionary processing technique showing the strict requirements for its printable raw materials, especially for starch-based materials. Here, they should have the ability to be easily extruded as "bio-ink", exhibiting favorable rheological behavior while holding their shape with/without a supporting microenvironment. Also, we will summarize the current two aspects of starch additive manufacturing: (1) the relationship between starch structure and its rheology and printability and (2) the application of starch-based 3D printing with its potential products.

10.2 PRINCIPLE OF ADDITIVE MANUFACTURING FOR STARCH

10.2.1 3D Printing Methods

Depending on the material property and forming principle, the main 3D printing methods can be divided into different modes such as extrusion-based printing (known as filament mode), jetting-based printing as well as solution shaping-based and powder shaping-based printing techniques (Ngo et al., 2018; Truby & Lewis, 2016). Table 10.1 presents a list of generally used 3D printing methods with a comparison of their feature sizes, print materials, and applications, including fused deposition modeling (FDM) (Liu et al., 2019; Han, Kundu, & Nag et al., 2019) and direct ink writing (DIW) (Truby and Lewis, 2016), highly adaptive for starch-based materials. Other methods include stereolithography (SLA) (Jiang et al., 2023), polyjet process, selective laser sintering (SLS) (Wang et al., 2023), 3D inkjet printing (Zhang, Moon & Ngo, 2020), digital light processing (DLP) (Tiller et al., 2019), and the materials of which contain starch or starch derivatives like oligosaccharide, cyclodextrin, etc., as the subsidiary materials for 3D printing.

The general principle of 3D printing has the following four main steps: (1) Obtaining 3D models. 3D modeling like toy vehicle as the edible starch-based product can be obtained either through computer 3D software design (e.g., CAD, SolidWorks, etc.) or by scanning the target real vehicle with a 3D dot-matrix scanner (scaled it down into a processable model using software such as Geomagic). (2) Slicing. Defined 3D models are sliced using software such as Cura, slic3r, etc., and the data for each layer is obtained and converted into G-code to control the travel of the printer (considering the adaptability of the model structure and filled starch-related materials). (3) Printing. The printer is set up according to the

TABLE 10.1 A brief summary of 3D printing methods

3D Printing Modes		Feature Size (μm)	Materials	Potential Applications
Extrusion- or filament-based	Fused deposition modeling (FDM)	200–400	Thermoplastic materials, e.g., PLA (polylactic acid), ABS (acrylonitrile butadiene styrene), nylon	Biomedical devices, sensors and actuators, soft robotics, composites, functional materials, lightweight materials, rapid prototyping and edible materials and food process
	Direct ink writing (DIW)	Depending on nozzle size	Biomaterials including starch, elastomers, metals and ceramics in micro/nanoparticle solutions, and thermosets	
Solution shaping-based	Stereolithography (SLA)	20–200	Laser curable polymers, such as GelMA (methylpropenylated gelatin), HAMA (methylpropenylated hyaluronic acid)	Microfluidic devices, architected materials, cell culture, regenerative medicine and high-resolution prototypes
	Direct light processing (DLP)	10–100		
	Two-photon lithography (TPL)	0.1–0.7		
Power shaping-based	Selective laser sintering (SLS)	100–400	Metals and alloys, ceramics, polymers, and semiconductors	Aerospace and automotive components, dentistry components, and building materials
	Direct metal laser sintering (DMLS)	50–100		
	Direct energy deposition (DED)	Relatively low		
Jetting-based	Material ink jetting	20–200	Low-viscous materials	Energy harvesting devices, multilayer structures, and flexible electronics and sensors, and some candies
	Aerosol jet printing (AJP)	10–200	Micro/nanoparticle solutions including starch solution, conductive pastes	
	Binder jetting		Ceramics and metals	

parameters for materials (taking note of the expansion characteristics of gelatinized starch), and they are linked together in a layer-by-layer fashion. (4) Post-processing. Post-processing such as de-supporting, sanding, and even heating for different goals is carried out to obtain the final products.

In detail, extrusion-based printing mainly including FDM and DIW is the most popular 3D printing technology for thermoplastic materials like acrylonitrile butadiene styrene (ABS) and polylactic acid (PLA) (Ngo et al., 2018) as well as starch (Chen et al., 2022; Sivamaruthi et al., 2022). The melted filament of FDM is extruded mechanically from the nozzle and printed onto the substrate (Figure 10.2a). The substrate of the build platform can also be controlled in a range of temperature for the cold-forming process. DIW is a mode of printing that is very similar to FDM except that the printing material used in the latter is generally semi-solid inks rather than filaments (Figure 10.2b) (Wan, Luo, & Liu, 2020). Unlike the high-temperature melting-solidification procedure of FDM, DIW can print starch-based materials in an ambient setting at room temperature (Cesarano, 1999; Zhang et al., 2022). Low-viscosity ink (e.g., low content of gelatinized starch) or shear-thinning ink (e.g., addition of starch nanoparticle as slip agent) are two different types of functional inks that can be

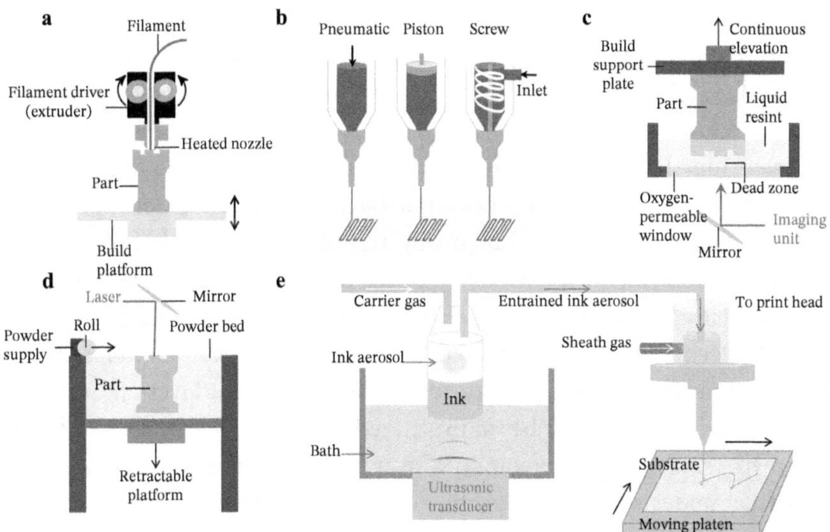

FIGURE 10.2 Working principles of common 3D printing methods. (a) Fused deposition modeling (FDM), (b) direct ink writing (DIW), (c) digital light processing (DLP), (d) selective laser sintering (SLS), and (e) aerosol jet printing (AJP).

used with DIW (such as printed all-starch tablets, see Section 10.4.4). Low-viscosity ink is easy to dispense from the nozzle, while shear-thinning ink necessitates a significantly stronger pushing force (Goh et al., 2021). Li et al. (2023) found that starch nanoparticles in thick 3D printing materials promote both shear-thinning and self-supporting properties (for details, see Section 10.4.2).

Solution shaping (or called liquid-based) 3D printing usually refers to the solidification and architecture formation of liquid resin-based materials, which is based on the principle of using a UV light source to treat UV-sensitive liquid resin layer by layer (Figure 10.2c). This mechanism has been used to develop a number of 3D printing techniques, including SLA, DLP, and mask stereolithography (MSLA/LCD) (Li et al., 2023), although there is still no starch-based materials used for these modes of 3D printing. When we put our eyes on their light-induced formation processes, SLA uses a point-source laser beam, whereas DLP accelerates the printing process by projecting pictures across the entire platform at each layer (Jiang et al., 2023). Both of them show high resolution and quick printing speed, but these processes have stringent requirements for the printability of the bioinks and that is why starch has not been used here yet. The bio-ink of liquid resin needs to be easily photo-crosslinked by exposure to light and have a low viscosity so that it can continually interface with the cured layers before curing (Li et al., 2023). In the future, there is also a chance for modified starch or nano-starch as additive materials to be used in liquid resin-based material.

Direct metal laser sintering (DMLS), SLS, and direct energy deposition are the three most popular powder-based 3D printing techniques. For these 3D printing, powdered printing ingredients are used. The powders are exposed to heat, typically from a powerful laser, which causes the powders to melt and bond together to form solid structures. More powders are cast over the previously printed layer after it has been completed, and the process is then repeated to produce the subsequent layer (Jiang et al., 2023). A high-energy laser beam is utilized in SLS to locally melt and weld the powders, creating 3D geometry layer by layer (Figure 10.2d). At present, these modes may not be useful to starch powder due to the overhigh energy-induced carbonization of starch materials.

Jetting-based printing has two categories, namely binder jetting and material jetting. The former is a mode of joining components that were previously in powder (e.g., ceramics and metals, as well as sugars and the potential materials of starch microgranules or nanoparticles) form by the deposition of a liquid-binding solution locally (Sun et al., 2015). Screen printing, 3D inkjet, and other processes that selectively deposit materials in droplet forms fall under the category of material jetting (Chen et al., 2020), including to some extent aerogel jet printing (AJP). AJP is one of the most recent and promising methods due to its high resolution (less than 10 μm), a wide range of ink viscosity (1–1000 cP), and capability of printing various types of materials (e.g., metals, ceramics, polymers, biomedical materials) (Jiang et al., 2023). In AJP, an atomizer (pneumatic or ultrasonic) is utilized to atomize the functional ink, and the generated aerosol of 2–5 μm (diameter) droplets is transported to the print head by a carrier gas flow (Zhang, Moon, & Ngo, 2020). Functionalized starch-enhanced gels are potential materials for AJP 3D printing.

10.2.2 Novel Concepts of Multidimensional Printing

With the development of 3D printing showing "materials-by-design" strategy, some emerging printing modes continue to enter the line of sight of researchers. Concepts and studies on multidimensional printing like 4D, 5D, and 6D have significantly increased during the past few years. In this section, we discuss the theories and studies behind cutting-edge additive manufacturing techniques for 4D, 5D, and 6D printing, as well as their prospective uses and difficulties for starch-related materials with their change of microstructure.

4D printing is an extension of 3D printing, and a "space-time axis" has been added to the 3D coordinate axis (Teng, Zhang, & Mujumdar, 2021). That means a physical or chemical change in the 3D-printed product over time (so-called the fourth dimension) when smart materials are integrated into them and respond to environmental or human-caused stimuli (such as water, temperature, pH, UV, and electrical and magnetic fields). When consumers consume 4D-printed things, their color, shape, and flavor may be stimulated to change, similar to a morphing starchy pasta with stresses introduced into a flat sheet for low-space packaging due to structural anisotropy or compositional heterogeneity (Tao et al., 2021). Ghazal et al. (2021) design a novel 4D food from potato starch, red cabbage juice, vanillin, and other fluids, simultaneously changing their color and flavor over time by external and internal pH stimuli. He, Zhang and Devahastin (2020) used a unique method for inducing spontaneous form change in 3D-printed purple sweet potato purees using microwave dehydration (Figure 10.3a). After spraying the 3D-printed gel with $NaHCO_3$ solution as a pH stimulus, Chen et al. (2021) used the starchy lotus root

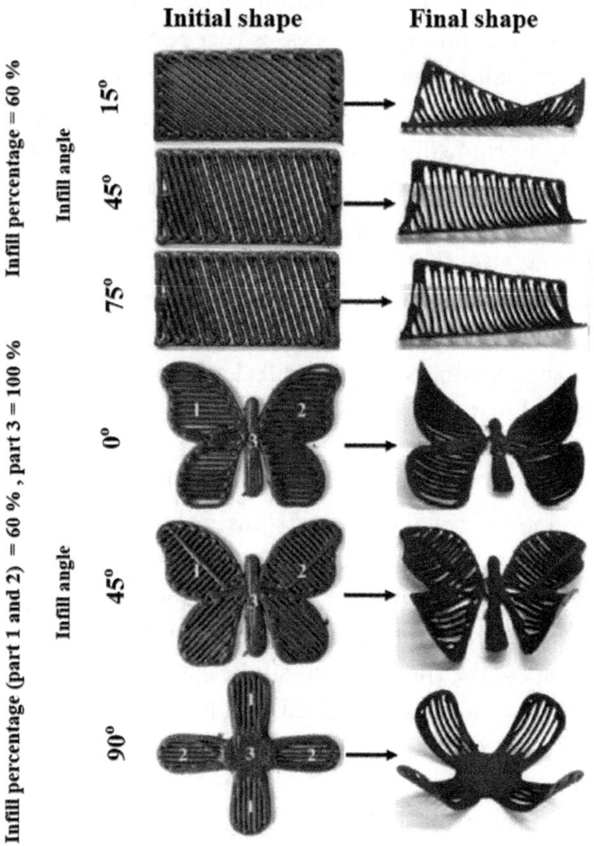

FIGURE 10.3 Application of 4D/5D food printing. (a) Microwave dehydration caused spontaneous shape alterations of printed samples. Reproduction from He, Zhang, and Devahastin (2020), with permission.

gel with the combination of anthocyanins and lemon yellow as a pH stimulus-responsive material to stimulate spontaneous color transformation (Figure 10.3d). Despite the advantages above, 4D printing has several drawbacks, including the absence of additional printing degrees of freedom and the requirement for support materials with lengthened processing time (Georgantzinos, Giannopoulos, & Bakalis, 2021). New additive manufacturing strategies like 5D and 6D printing have also emerged and developed with additional capabilities and certain advantages over 4D printing (Ghazal et al., 2023).

The idea of 5D printing was firstly proposed in 2016, and the research team at Mitsubishi Electric Research Laboratories (MERL) was the first to use this cutting-edge technology, thanks to William Yerazunis (He, Zhang, & Fang, 2020). The main benefit of using 5D printing technology is the capacity to print in five separate axes, which improves the strength of any necessary layers, planes or curves, producing robust things with many dimensions. In the starch process, 5D printing, which has not been used yet, may create the most intricate, and curved food structures (e.g., morphing pasta as shown before) with less material and greater strength to preserve their characteristic shape during shipping. Additionally, 5D printing makes it possible to print goods with any infill type and infill density theoretically while still preserving the product's shape after printing (Ghazal et al., 2023). This makes it possible to manufacture products with a variety of textures to suit a variety of consumer preferences. The 6D printing may be thought of as the product of a five-axis printer that use smart materials under stimulating conditions. Compared to 5D printing, 6D printing combines the capacity to produce the most complex structures with increased strength and the need for less material. However, to the best of our knowledge, 6D printing technology hasn't yet been used in starch applications and any other industries. Future research could lead to the discovery of fresh smart materials (added into starch-based materials or prepared from modified starches), stimuli, and mechanisms to achieve desired changes in various attributes and make significant advancements in 4D/5D/6D printing scenarios.

10.2.3 Typical 3D Printing for Potential Starch Process

Due to the distinctive physicochemical properties of starches, as well as their resources and costs, a large number of them are appropriate for food and non-food printing applications. The most typical approach utilized in commercial starch 3D printers is the material extrusion technique, especially for new food design. Furthermore, the use of powder bed fusion, material jetting, and binder jetting to create specific types of starch and starch derivatives have also been studied (Enfield et al., 2022). Here, we provide a brief overview of four different 3D printing technologies, describing their possible use in the production of 3D products (food cases of starch in this section).

Extrusion-based printing is the most studied 3D printing technology for starch printing because it allows for the use of a wide range of food ingredients as suitable inks. Hydrogels (including starch-based or starch-enhanced types), sugar frostings (or starch degradation products), as well as cheeses and chocolate are the most explored inks due to their ability to be extruded easily through the syringe and keep their shape after printing (Outrequin et al., 2023). The texture and nutritional characteristics of printed foods can be customized using these starch-containing materials, sometimes by cooling or heating to obtain the final products. Starch foods with relatively basic structures, such snacks and desserts, can easily be made using extrusion-based printing, but it is more challenging (see Section 10.4) to make them with complicated structures to mimic the texture of meat, fish, fruits or vegetables (Keerthana et al., 2020; Derossi et al., 2018; Lille et al., 2018). For undesirable texture of food products, it is usually required to add ingredients that enhance their textural characteristics. For example, adding more potato starch to the recipe resulted in mashed potatoes with good textures (Liu et al., 2018).

Binder jetting is a repetitive procedure that involves adding a thin layer (overlapping) of food powder and binding a powdered substance using a drop-on-demand liquid binder to create a 3D object (Godoi, Prakash, & Bhandari, 2016). Usually, binder jetted materials have poor mechanical properties because of the formation of weak bindings between the particles. In order to strengthen their mechanical strength, the as-prepared objects frequently need to be processed thereafter (Enfield et al., 2022). Binding powdered substances like sugar is possible with the help of flavor liquids and colors (Sher & Tutó, 2015). It is possible to create distinctive, intricate, and flavorful confectionary goods with this technique, and part of the ingredients could be instead of starch or dextrin. Complex designs used to take a long time to construct (Mantihal, Kobun, & Lee, 2020), but printing efficiency for 3D products has decreased. According to Holland et al. (2018), edible goals have been made using the technique of mixing semi-crystalline cellulose powder with a xanthan gum solution as a liquid binder. The high crystallinity of resistant starch powder could also be considered for mixing as 3D printing materials.

Foods can be printed using material jetting, which is commonly done using either drop-on-demand printing or continuous jet printing. It is difficult to retain the final 3D structure of the food object since the material jetting approach only print materials with low viscosity like dilute starch solution. Consequently, it is mostly employed to create 2D graphics on the surfaces of products, such as printing on a cake or patterns on edible films or coatings. The printing parameters have a significant impact on the printing precision and accuracy of the objects (Enfield et al., 2022), as well as the compatibility of the starch ink with the substrate surface, and the viscosity and other rheological properties of the starch ink (see below). Various starch-containing food products, such as doughs and batters, as well as other materials like chocolate, butter, cream, sauces, purees, jams, and jellies, are used as food inks in a commercial printer called "FoodJet".

10.3 STRUCTURAL CONTROL OF PRINTABLE STARCH

10.3.1 Structural Modification

Due to its excellent adaptability and capacity, starch is a widely utilized biopolymer to be combined with other components to produce a paste-like material appropriate for 3D printing (Chen et al., 2022; Zhuang, Greenberg, & He, 2021). The thermal, rheological, and surface properties of starch materials have effects on the printing precision and form stability of 3D-printed objects. According to Gulzar et al. (2023), the desirable properties of starches are summarized as low gelatinization temperature, high gel swelling ability or sometimes high solubility, good pasting (shear-thinning with high recovery), and the capacity to form powerful hydrogels. Low adhesiveness and pasting viscosity are also to some extent needed for 3D-printing applications.

A variety of physical or chemical modifications to improve the structure and functional properties of starch to make it more suitable for 3D printing, especially for starch gel-forming or starch-induced crosslinking. As shown in Table 10.2, physical action alters the starch microstructure via green modification, whereas chemical modification involves the introduction of chemical agents to provide new functional groups onto starch molecules (Nawaz et al., 2020). The results indicate that physical methods such as high-pressure processing (Larrea-Wachtendorff, Del Grosso, & Ferrari, 2022), pulsed electric field (Maniglia et al., 2021), ultrasonication (Xu, Zhang, & Bhandari, 2020), and microwave heating (Oyeyinka et al., 2021) (which have been comprehensively discussed for their modifications on the hierarchical structure of starch in the chapters above) can significantly enhance the rheological and mechanical properties of starch-based materials for 3D printing. Additionally, starches can be pre-treated with cold plasma (Okyere, Rajendran, & Annor, 2022), ozone (Maniglia et al., 2019), and enzymatic (Park, Kang, & Rho et al., 2020) processes to give them the right flow behavior, high resolution, and adequate strength for 3D printing.

The process of modifying starch utilizing fresh, developing technologies is still in its infancy for the diverse needs of 3D printing. Except for the

TABLE 10.2 Some Cases of Modification Method for Improving the 3D Printing Ability of Starch

Physical Pre-Treatment	Starch Type	Processing and Results	References
High-pressure processing (HPP)	Potato starch	HPP for 15 minutes at 25°C, 40°C and 50°C at 600 MPa (higher viscosity and higher modulus)	Larrea-Wachtendorff, Tabilo-Munizaga, & Ferrari (2019)
Pulsed electric field (PEF)	Wheat and cassava starch	**PEF in three conditions:** (1) 15 kV/cm, 25 kJ/kg; (2) 25 kV/cm, 25 kJ/kg; (3) 25 kV/cm, 50 kJ/kg (the best sample for 3D printing)	Maniglia et al. (2021)
Ultrasonic-microwave (UM)	Wheat starch	UM at power 80 W (for good printing performance)	Xu, Zhang, & Bhandari (2020)
Cold plasma (CP)	Corn starch	CP for 30 minutes at 400–800 W (with decreased peak, final, and setback viscosities of starch samples)	Wu et al. (2019)
Ozone attacking (OA)	Cassava starch	OA for 30 minutes at 43 mg O_3L^{-1} (modified starch has good printability)	Maniglia et al. (2019b)
Enzymatic hydrolysis (EH)	Rice starch	4-α-GTase modification of starch changed its gelation characteristics and reduced its viscosity	Park, Kang, Rho, & Kim (2020b)

structural modification of starch itself, the addition of hydrophilic colloids and other chemical components has proven to be a successful method for addressing the structure, texture, quality, functional (for non-food products), and nutritional inadequacies (for food products) created by starch as a single component of bio-ink (Chen et al., 2022). Comparative studies between different methods can also be carried out to evaluate their viabilities as a commercially viable technology for starch modification (Gulzar et al., 2023).

10.3.2 Rheology and Printability

Extrusion 3D printing is currently the most widely used and studied in starch additive manufacturing, which is focused here with the emphasis on the key factors of rheology and printability (Figure 10.4). A material is considered printable for extrusion-based 3D printing when both of the following requirements are met: (1) the extruded materials must be able to hold the intended shape; (2) the extrudate filament must be extruded continuously; and (3) the finished construction must also be robust (so-called self-supporting) (Duty et al., 2018). Previously, "printability" was described as the ratio of the height to the desired height of printed products (In et al., 2021). However, the extrudability and shape-holding capabilities of the printable inks are not included in this criterion. Therefore, it is crucial to remember that the fidelity of the printed 3D product to the model is equally important (Outrequin et al., 2023), especially for the starches with difficult to be precisely controlled due to their expansive or fluidizable states at different concentrations and gelatinization degrees.

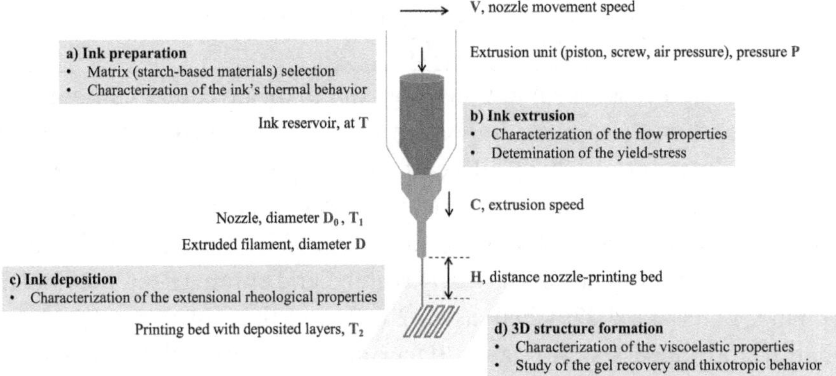

FIGURE 10.4 Extrusion 3D printing-related parameters schematic diagram.

As is known to all, rheology has become an important tool for judging whether materials are printable under environmental conditions. Added inks often need the right rheology qualities to ensure a smooth printing process and high-quality prints in order to be extruded and shaped successfully (Cheng et al., 2022). During the printing process, the rheological factors like yield stress (τ_{yield}), consistency (*n*), and flow behavior (*K*) indices all affect how easily ink flows, which has a direct effect on the necessary pressure for extrudability and ensuing printing. As shown above, the creation of the self-supported 3D item also requires proper stability of the stereostructure. Thus, to build the 3D starch products, it is important to understand the viscoelastic moduli, the relaxation time (λ), and the extensional properties of their inks. Besides, it is important to note that when starch material is being processed, the temperature can be used to control how the object structure is created and shaped, for instance, by using thermosensitive biopolymers (like modified starch), whose properties can change depending on the temperature from being a viscous liquid to a gel or vice versa.

In order to achieve good printability, knowledge of the optimal printing starch material qualities alone is insufficient; the appropriate adjusting of the printer settings should also be considered (Yuk & Zhao, 2018). According to Outrequin et al. (2023), these related print parameters (as shown in Figure 10.4) include the nozzle speed (*V*), the extrusion pressure (*P*), the extrusion speed (*C*), the temperature of the ink in the reservoir (*T*), the temperature of the printing nozzle (T_1), the temperature of the printing bed (T_2), the nozzle diameter (D_0), and the distance between the nozzle and the printing bed (*H*).

10.4 FOOD AND NON-FOOD APPLICATIONS OF STARCH 3D PRINTING

Starch has been used in many fields as the main or added components (starch concentration as an important factor) of the printing substrates (Zhang, Zheng, et al., 2022). Recently, the popular application areas of starch-containing 3D printing include food, medicine, environment, packaging industries, and so on (Liu et al., 2023). General starch-based food by 3D printing has been widely studied and summarized previously (Zhang, Li, et al., 2022; Chen et al., 2022), and thus, in this section, we mainly introduce starch-related 3D printing of personalized nutrition foods and plant meat products, starch-based films, starch hydrogel used in biomedicine, etc.

10.4.1 Personalized Nutrition Foods

Personalized nutrition refers to the production of food that is adapted to the preferences and needs of customers, which may be tailored to fulfill the demands of various sectors, demographic groups, and ages (Derossi et al., 2018). According to the differences in individual needs, 3D printing is expected to alter the amount and type of nutrients for personality (Zhang, Zheng, et al., 2022). For example, people with diabetes should avoid starchy foods because of their high sugar content, while 3D-printed starch-based foods can offer tailored meals for them (Portanguen et al., 2019). Another case is that a person of advanced age has a difficulty to swallow and digest food, which not only lowers the quality of life but also increases the risk of malnutrition and other problems (Alagiakrishnan, Bhanji, & Kurian, 2013), and the strategy of 3D printing for producing easy-to-swallow food is promising. Customers with nutritional shortages may explore nutritional fortification of the 3D-printed material with a balanced mix of nutrients, like children who need customized dosage with an alternative oral form (i.e., gummy) with eye-catching appearance and appropriate organoleptic characteristics (Figure 10.5) (Herrada-Manchón et al., 2020). Also, nutrient-dense 3D-printed snacks have been investigated using high-fiber, high-protein composite flour as bio-ink based on extrusion 3D printers (Krishnaraj et al., 2019). 3D-printed fiber-rich snack food was developed by Keerthana et al. (2020) using fiber-rich mushroom powder with wheat, which increased its viscosity to create a supporting framework. Furthermore, starch can be printed and used as the carrier of micronutrients (like catechin, procyanidin, etc.), for potential stabilization and precision control of their distribution and release in human digestive systems (Zeng et al., 2021).

10.4.2 Starch-Based Meat Analogues

Based on health, environmental, and animal welfare considerations, people are increasingly inclined to semi-vegetarian, vegan, or "flexible" consumption patterns. It has been demonstrated that eating regimens low in meat (such as vegetarian or vegan diets) or high in protein replacement (such as cell culture and plant-based protein replacement) expand slowly (OECD & FAO, 2021). Since starch is a source of carbs (one of the most important components in diet) and shows the performance of 3D printing with/without modification, several researchers have used it in the creation of 3D-printed resemble meat. Ko et al. (2021) used a coaxial nozzle-assisted 3D food printer to print potato starch and soy protein

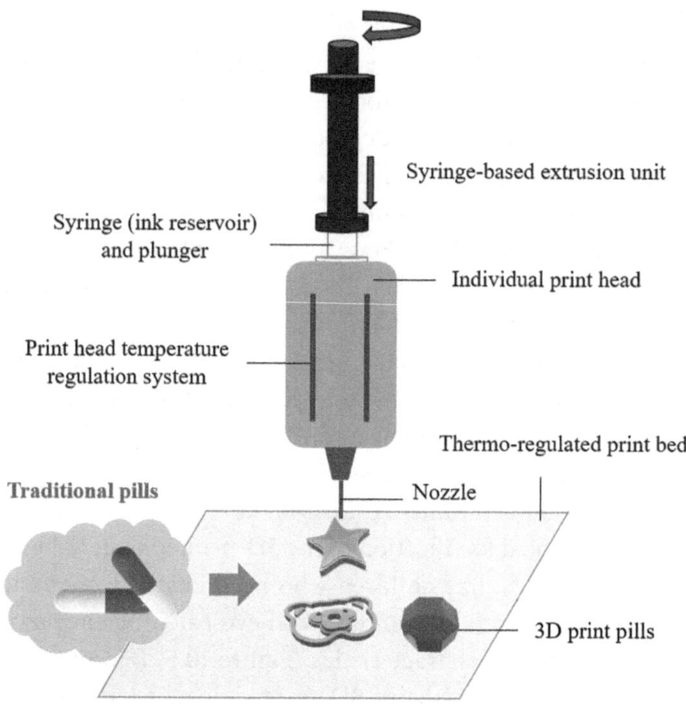

FIGURE 10.5 3D print gummy instead of pills.

isolate into a 3D-printed product with a fibrous texture that is close to the hardness of real beef. Shi et al. (2023) used rice starch, soy protein isolate, and xanthan gum as food ink to adjust the geometry by improving the precision of printing filament (~100 μm) for the preparation of plant-based fish analogues (Figure 10.6). Nanosized starch was used with lutein to enhance the printability of surimi to prepare antioxidant functional surimi products (Li et al., 2023). Also, Li et al. (2023) used nano-starch (approximately 2%) to enhance the rheology and printability of carrageenan emulsion gel to prepare high-fidelity and nutrition-fortified fish fat mimics, which has a tightly bound water composition similar to real fish tissue (Figure 10.7).

10.4.3 Starch-Based Films

Biodegradable films can also be created using 3D printing by ingredients based on starch. A more sustainable food system may result from 3D-printed starch-based films since they can be used to create novel food packaging modes reducing food and agricultural waste (Chen et al., 2022). For instance, by combining starch and agricultural waste, 3D printing was

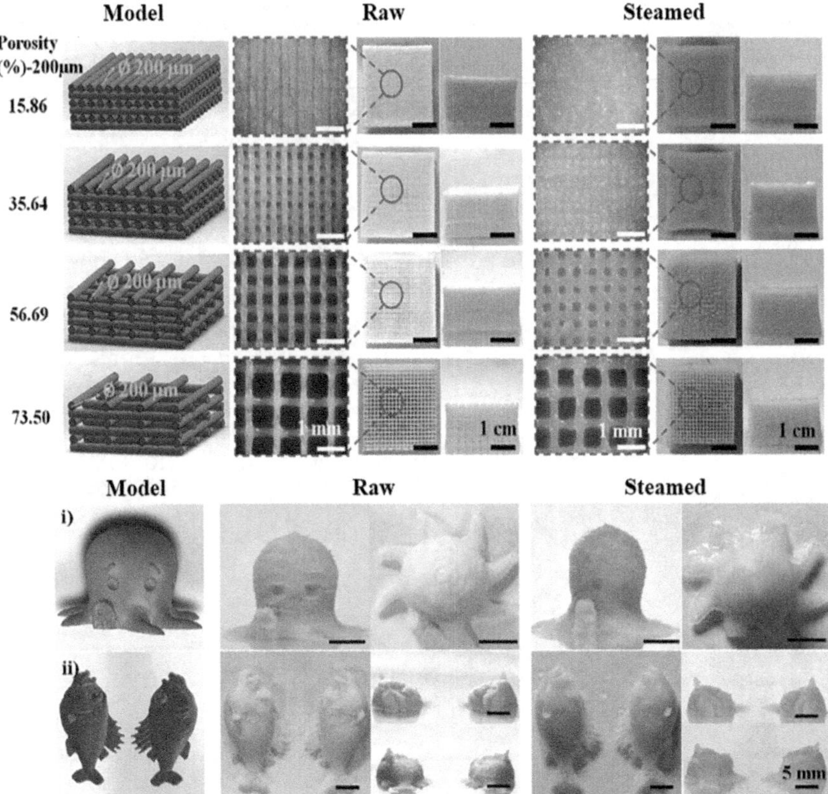

FIGURE 10.6 High-precision fish simulators were prepared by using rice starch. Reproduction from Shi et al. (2023), with permission.

utilized to produce biodegradable films that can replace petroleum-based plastics (Scaffaro et al., 2022). Maize starch-gelatin sheets with strong mechanical and antibacterial properties have been made using 3D printing, making them suitable as active packaging materials (Leaw, Kong, & Pui, 2021).

In addition, 3D-printed films based on starch are produced with intelligent and active functions. For instance, sandwich-type indicator films have been constructed utilizing 3D printing technology, and the formula includes chitosan (as a film-forming substrate), mulberry anthocyanin (as a colorimetric sensor), lemongrass essential oil (as antibacterial and antioxidant), and cassava starch (as a protective layer) (Li et al., 2022). It demonstrated how to increase the freshness of cold meat and check its freshness by analyzing color variations in the indicators brought on by pH fluctuations.

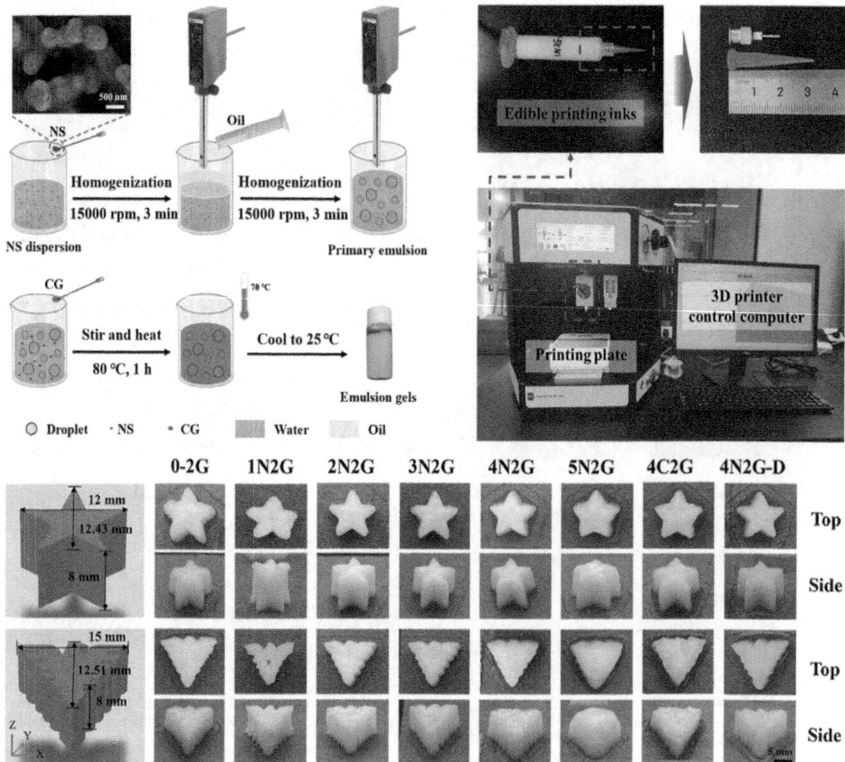

FIGURE 10.7 High-fidelity and nutrition-fortified fish fat mimics were prepared by using the printability of nano-starch-enhanced emulsion gel. Reproduction from Li et al. (2023), with permission.

10.4.4 Starch Hydrogel in Biomedicine

3D bioprinting starch hydrogels as natural biomaterials have gradually been used in biomedicine such as tissue engineering and drug delivery (Sivamaruthi et al., 2022; Lam et al., 2002). For example, Zhuang et al. (2021) developed a novel 3D printable starch nanocomposite hydrogel with highly enhanced biocompatibility, which promotes 3D cell growth by adding gelatin nanoparticles and collagen. 3D bioprinting starch/gelatin gum and starch/chitosan scaffolds have been studied for Schwaan cell and neural cell growth, respectively (Butler et al., 2020; Zhang et al., 2021). Ng et al. (2023) successfully incorporated pre-gummed starch (PGF) and glycerol as molecular lubricants and hydrophilic plasticizers into adipose-derived stem cells-laden IPN hydrogel formulations. This novel glycerol PGF-gelatin gum composite hydrogel has excellent shear recovery ability, improved resistance to step strain deformation, and sufficient stiffness

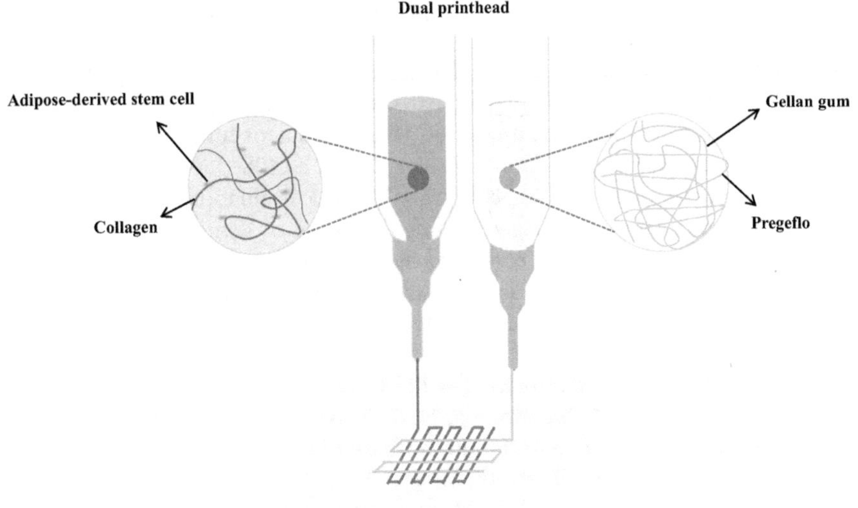

FIGURE 10.8 Incorporation of pregelatinized starch enables extrusion-based 3D bioprinting of cell-laden hydrogels.

to serve as a favorable environment for cells (Figure 10.8). Furthermore, starch could be added into N,O-carboxymethyl chitosan and printed for wound dressing applications, like loading mupirocin for quantified release (Naseri et al., 2021). Zheng et al. (2023) also designed an oxidized starch hydrogel loaded with coagulation factor Ca^{2+} ions (CaOMS), with the advantages of rapid and effective hemostasis, effective wound healing, and adaptability to fit various wound shapes.

For target delivery, 3D printing of customized starch tablets with different geometries (i.e. cylinders and pyramids) has been investigated for the release kinetics of hydrophobic drugs (González et al., 2022). Bom et al. (2020) incorporated starch into alginate-based 3D hydrogel patches for drug delivery and explored their physicochemical and release kinetics, to improve the properties of the drug delivery system while maintaining the desired quality characteristics. The studies above provide an idea for the future drug release regulation strategy. Also, Cardoso et al. (2023) confirmed that the addition of starch (~13%) to chitosan hydrogels improved the structural properties of chitosan-based scaffolds and designed scaffold systems with tailored drug delivery properties. Chatzitaki et al. (2022) used semi-solid extrusion 3D printing to manufacture starch-based ISO soft dosage forms, which is a viable way to achieve pediatric-friendly formulations in resource-poor environments.

REFERENCE, BIBLIOGRAPHY OR WORKS CITED

Aikaterini, T. C., Emmanouela, M., Nikolaos, B., Christos, R., Christina, K., & Dimitrios, G. F. 2022. Semi-solid extrusion 3D printing of starch-based soft dosage forms for the treatment of paediatric latent tuberculosis infection. *Journal of Pharmacy and Pharmacology* 74 (10): 1498–1506.

Alagiakrishnan, K., Bhanji, R. A., & Kurian, M. 2013. Evaluation and management of oropharyngeal dysphagia in different types of dementia: A systematic review. *Archives of Gerontology and Geriatrics* 56 (1): 1–9.

Bom, S., Catarina, S., Rita, B., Ana, M. M., Patrizia, P., Ricardo, C., Pedro, C. P., Helena, M. R., & Joana, M. 2020. Effects of starch incorporation on the physicochemical properties and release kinetics of a-based 3D hydrogel patches for topical delivery. *Pharmaceutics* 12 (8): 719.

Butler, H. M., Naseri, E., MacDonald, D. S., Tasker, R. A., & Ahmadi, A. 2020. Optimization of starch-and chitosan-based bio-inks for 3D bioprinting of scaffolds for neural cell growth. *Materialia* 12: 100737.

Cardoso, S., Francisco, N., Nuno, M., Ana, B., & Isabel, A. C. Ribeiro. 2023. Improving chitosan hydrogels printability: A comprehensive study on printing scaffolds for customized drug delivery. *International Journal of Molecular Sciences* 24 (2): 973.

Cesarano, J., Dimos, I. D., Danforth, S. C., & Cima, M. J. 1999. A review of robocasting technology. *Solid Freeform and Additive Fabrication* 542: 133–139.

Chen, C. L., Jiang, J. M., He, W. J., Lei, W., Hao, Q. L., & Zhang, X. G. 2020. 3D printed high-loading lithium-sulfur battery toward wearable energy storage. *Advanced Functional Materials* 30 (10): 1909469.

Chen, C., Zhang, M., Mujumdar, A. S., & Phuhongsung, P. 2021. Investigation of 4D printing of lotus root-compound pigment gel: Effect of pH on rapid colour change. *Food Research International* 148:110630.

Chen, H., Xie, F., Chen, L., & Zheng, B. 2019. Effect of rheological properties of potato, rice and corn starches on their hot-extrusion 3D printing behaviors. *Journal of Food Engineering* 244: 150–158.

Chen, Y., McClements, D. J., Peng, X., Chen, L., Xu, Z., Meng, M., Zhou, X., Zhao, J., & Jin, Z. 2022. Starch as edible ink in 3D printing for food applications: a review. *Critical Reviews in Food Science and Nutrition* 64 (2): 1–16.

Cheng, Y., Yu, F., Liang, M., Pei, L. Y., Dusan, L., Hongxia, W., & Zhang, Y. 2022. Rheology of edible food inks from 2D/3D/4D printing, and its role in future 5D/6D printing. *Food Hydrocolloids* 132 (6): 107855

Derossi, A., Caporizzi, R., Azzollini, D., & Severini, C. 2018. Application of 3D printing for customized food. A case on the development of a fruit-based snack for children. *Journal of Food Engineering* 220: 65–75.

Duty, C., Ajinjeru, C., Kishore, V., Compton, B., Hmeidat, N., Chen, X., Liu, P., Hassen, A. A., Lindahl, J., & Kunc, V. 2018. What makes a material printable? A viscoelastic model for extrusion-based 3D printing of polymers. *Journal of Manufacturing Processes* 35: 526–537.

Enfield, R. E., Pandya, J. K., Lu, J., McClements, D. J., & Kinchla, A. J. 2022. The future of 3D food printing: Opportunities for space applications. *Critical Reviews in Food Science and Nutrition* 63 (29): 10079–10092.

Georgantzinos, S. K., Giannopoulos, G. I., & Bakalis, P. A. 2021. Additive manufacturing for effective smart structures: The idea of 6D printing. *Journal of Composites Science* 5 (5): 119.

Ghazal, A. F., Zhang, M., Bhandari, B., & Chen, H. 2021. Investigation on spontaneous 4D changes in color and flavor of healthy 3D printed food materials over time in response to external or internal pH stimulus. *Food Research International* 142: 110215.

Ghazal, A. F., Zhang, M., Arun, S. M., & Mohamed, G. 2023. Progress in 4D/5D/6D printing of foods: Applications and R&D opportunities. *Critical Reviews in Food Science and Nutrition* 63 (25): 7399–7422.

Godoi, F. C., Prakash, S., & Bhandari, B. R. 2016. 3D printing technologies applied for food design: Status and prospects. *Journal of Food Engineering* 179: 44–54.

Goh, G. L., Zhang, H., Chong, T. H., & Yeong, W. Y. 2021. 3D Printing of multilayered and multimaterial electronics: A review. *Advanced Electronic Materials* 7 (10): 2100445.

González, K., Larraza, I., Berra, G., Eceiza, A., & Gabilondo, N. 2022. 3D printing of customized all-starch tablets with combined release kinetics. *International Journal of Pharmaceutics* 622: 121872.

Gulzar, S., Narciso, J. O., Elez-Martínez, P., Martín-Belloso, O., & Soliva-Fortuny, R. 2023. Recent developments in the application of novel technologies for the modification of starch in light of 3D food printing. *Current Opinion in Food Science* 52: 101067.

Han, T., Kundu, S., Nag, A., & Xu, Y. 2019. 3D printed sensors for biomedical applications: A review. *Sensors* 19 (7): 1706.

He, C., Zhang, M., & Devahastin, S. 2020. Investigation on spontaneous shape change of 4D printed starch-based purees from purple sweet potatoes aAs induced by microwave dehydration. *ACS Applied Materials & Interfaces* 12 (34): 37896–37905.

He, C., Zhang, M., & Fang, Z. 2020. 3D printing of food: pretreatment and post-treatment of materials. *Critical Reviews in Food Science and Nutrition* 60 (14): 2379–2392.

Herrada-Manchón, H., Rodríguez-González, D., Fernández, M. A., Suñé-Pou, M., Pérez-Lozano, P., García-Montoya, E., & Aguilar, E. 2020. 3D printed gummies: Personalized drug dosage in a safe and appealing way. *International Journal of Pharmaceutics* 587: 119687.

Holland, S., Foster, T., MacNaughtan, W., & Tuck, C. 2018. Design and characterisation of food grade powders and inks for microstructure control using 3D printing. *Journal of Food Engineering* 220: 12–19.

In, J., Jeong, H., Song, S., & Min, S. C. 2021. Determination of material requirements for 3D gel food printing using a fused deposition modeling 3D printer. *Foods* 10 (10): 2272.

Jandyal, A., Chaturvedi, I., Wazir, I., Raina, A., & Mir Irfan Ul Haq. 2022. 3D printing - A review of processes, materials and applications in industry 4.0. *Sustainable Operations and Computers* 3: 33–42.

Ji, S. Y., Xu, T., Li, Y., Li, H. Y., Zhong, Y. H., & Lu, B. Y. 2022. Effect of starch molecular structure on precision and texture properties of 3D printed products. *Food Hydrocolloids* 125:107387.

Jiang, Y., Islam, M. N., He, R., Huang, X., Cao, P. F., Advincula, R. C., Dahotre, N., Dong, P., Wu, H. F., & Choi, W. 2023. Recent advances in 3D printed sensors: Materials, design, and manufacturing. Advanced Materials Technologies 8 (2): 2200492.

Keerthana, K., Anukiruthika, T., Moses, J. A., & Anandharamakrishnan, C. 2020. Development of fiber-enriched 3D printed snacks from alternative foods: A study on button mushroom. *Journal of Food Engineering* 287: 110116.

Kim, H. W., & Rhee, M. S. 2020. Space food and bacterial infections: Realities of the risk and role of science. *Trends in Food Science & Technology* 106: 275–287.

Ko, H. J., Wen, Y., Choi, J. H., Park, B. R., Kim, H. W., & Park, H. J. 2021. Meat analog production through artificial muscle fiber insertion using coaxial nozzle-assisted three-dimensional food printing. *Food Hydrocolloids* 120: 106898.

Krishnaraj, P., Anukiruthika, T., Choudhary, P., Moses, J. A., & Anandharamakrishnan, C. 2019. 3D extrusion printing and post-processing of fibre-rich snack from indigenous composite flour. *Food and Bioprocess Technology* 12(10): 1776–1786.

Lam, C.X.F., X.M Mo., S.H Teoh., & D.W Hutmacher. 2002. Scaffold development using 3D printing with a starch-based polymer. *Materials Science and Engineering* 20 (1-2): 49–56.

Larrea-Wachtendorff, D., Del Grosso, V., & Ferrari, G. 2022. Evaluation of the physical stability of starch-based hydrogels produced by high-pressure processing (HPP). *Gels* 8 (3): 152.

Larrea-Wachtendorff, D., Tabilo-Munizaga, G., & Ferrari, G. 2019. Potato starch hydrogels produced by high hydrostatic pressure (HHP): A first approach. *Polymers* 11 (10): 1673

Leaw, Z. E., Kong, I., & Pui, L. P. 2021. 3D printed corn starch-gelatin film with glycerol and hawthorn berry (Crataegus pinnatifida) extract. *Journal of Food Processing and Preservation* 45 (9): 15752e

Li, G., Zhan, J., Hu, Z., Huang, J., Yao, Q., Yuan, C., Chen, J., & Hu, Y. 2023. Effects of nano starch-lutein on 3D printing properties of functional surimi: Enhancement mechanism of printing effects and anti-oxidation. *Journal of Food Engineering* 346: 111431.

Li, J., Niu, R., Zhu, Q., Yao, S., Zhou, J., Wang, W., Chen, Q., Yin, J., Liu, D., & Xu, E. 2023. Nanostarch-enhanced 3D printability of carrageenan emulsion gel for high-fidelity and nutrition-fortified fish fat mimics. *Food Hydrocolloids* 145: 109099.

Li, S., Jiang, Y., Zhou, Y., Li, R., Jiang, Y., Alomgir Hossen, M., Dai, J., Qin, W., & Liu, Y. 2022. Facile fabrication of sandwich-like anthocyanin/chitosan/lemongrass essential oil films via 3D printing for intelligent evaluation of pork freshness. *Food Chemistry* 370: 131082.

Li, W., Wang, M., Ma, H., Chapa-Villarreal, F. A., Lobo, A. O., & Zhang, Y. S. 2023. Stereolithography apparatus and digital light processing-based 3D bioprinting for tissue fabrication. *Iscience* 26 (2): 106039.

Lille, M., Nurmela, A., Nordlund, E., Metsä-Kortelainen, S., & Sozer, N. 2018. Applicability of protein and fiber-rich food materials in extrusion-based 3D printing. *Journal of Food Engineering* 220: 20–27.

Lipton, J., Arnold, D., Nig, F., Lopez, N., Cohen, D., & Lipson, N. N. H. 2010. *Multi-material food printing with complex internal structure suitable for conventional post-processing*. Austin: University of Texas at Austin.

Liu, W. M., Chen, L., McClements, D. J., Peng, X. W., & Jin, Z. Y. 2023. Recent trends of 3D printing based on starch-hydrocolloid in food, biomedicine and environment. *Critical Reviews in Food Science and Nutrition* 1–15.

Liu, Z., Wang, Y., Wu, B., Cui, C., Guo, Y., & Yan, C. 2019. A critical review of fused deposition modeling 3D printing technology in manufacturing polylactic acid parts. *The International Journal of Advanced Manufacturing Technology* 102 (9-12): 2877–2889.

Liu, Z., Zhang, M., Bhandari, B., & Wang, Y. 2017. 3D printing: Printing precision and application in food sector. *Trends in Food Science & Technology* 69: 83–94.

Liu, Z., Zhang, M., Bhandari, B., & Yang, C. 2018. Impact of rheological properties of mashed potatoes on 3D printing. *Journal of Food Engineering* 220: 76–82.

Lorenz, T., Iskandar, M. M., Baeghbali, V., Ngadi, M. O., & Kubow, S. 2022. 3D food printing applications related to dysphagia: A narrative review. *Foods* 11 (12): 1789.

Manapat, J. Z., Chen, Q., Ye, P., & Advincula, R. C. 2017. 3D Printing of polymer nanocomposites via stereolithography. *Macromolecular Materials and Engineering* 302 (9): 1600553.

Maniglia, B. C., Lima, D. C., Matta Junior, M. D., Le-Bail, P., Le-Bail, A., & Augusto, P. E. D. 2019. Hydrogels based on ozonated cassava starch: Effect of ozone processing and gelatinization conditions on enhancing 3D-printing applications. *International Journal of Biological Macromolecules* 138: 1087–1097.

Maniglia, B. C., Pataro, G., Ferrari, G., Augusto, P. E. D., Le-Bail, P., & Le-Bail, A. 2021. Pulsed electric fields (PEF) treatment to enhance starch 3D printing application: Effect on structure, properties, and functionality of wheat and cassava starches. *Innovative Food Science & Emerging Technologies* 68: 102602.

Mantihal, S., Kobun, R., & Lee, B. 2020. 3D food printing of as the new way of preparing food: A review. *International Journal of Gastronomy and Food Science* 22: 100260.

Molitch-Hou, M. 2020. *Chocolate 3D printing with mass customization around the corner, says FoodJet*. (Vol. 2023). 3DPrint.Com. https://www.foodjet.com/chocolate-3d-printing.

Naseri, E., Cartmell, C., Saab, M., Kerr, R. G., & Ahmadi, A. (2021). Development of N,O-carboxymethyl Chitosan-starch biomaterial inks for 3D printed wound dressing applications. *Macromolecular Bioscience* 21 (12): 2100368.

Nawaz, M., Ali, Q., Azeem, M., Asghar, F., Ishaq, H., & Iqbal, N. 2020. Capparis decidua Edgew (Forssk.) stem extract alleviates the water stress perturbations in wheat (Triticum aestivum L.) at early growth stage. *Journal of Bioresource Management* 7 (4): 13.

Ng, J. Y., Yu, P Y., Murali, D. M., Liu, Y. S., Gokhale, R., & Ee, P. L. R. 2023. The influence of pregelatinized starch on the rheology of a gellan gum-collagen IPN hydrogel for 3D bioprinting. *Chemical Engineering Research and Design* 192: 477–86.

Ngo, T. D., Kashani, A., Imbalzano, G., Nguyen, K. T. Q., & Hui, D. 2018. Additive aanufacturing (3D Printing): A review of materials, methods, applications and challenges. *Composites Part B: Engineering* 143: 172–96.

Obrist, M., Tu, Y., Yao, L., & Velasco, C. 2019. Space food experiences: Designing passenger's eating experiences for future space travel scenarios. *Frontiers in Computer Science* 1.

OECD, & FAO. 2021. OECD-FAO Agricultural Outlook 2021–2030. Paris: OECD Publishing

Okyere, A. Y., Rajendran, S., & Annor, G. A. 2022. Cold plasma technologies: Their effect on starch properties and industrial scale-up for starch modification. *Current Research in Food Science* 5: 451–463.

Outrequin, T. C. R., Gamonpilas, C., Siriwatwechakul, W., & Sreearunothai, P. 2023. Extrusion-based 3D printing of food biopolymers: A highlight on the important rheological parameters to reach printability. *Journal of Food Engineering* 342: 111371.

Oyeyinka, S. A., Akintayo, O. A., Adebo, O. A., Kayitesi, E., & Njobeh, P. B. 2021. A review on the physicochemical properties of starches modified by microwave alone and in combination with other methods. *International Journal of Biological Macromolecules* 176: 87–95.

Park, H. R., Kang, J., Rho, S., & Kim, Y. 2020. Structural and physicochemical properties of enzymatically modified rice starch as influenced by the degree of enzyme treatment. *Journal of Carbohydrate Chemistry* 39 (5): 250–266.

Phuhongsung, P., Zhang, M., & Bhandari, B. 2020. 4D printing of products based on soy protein isolate via microwave heating for flavor development. *Food Research International* 137: 109605.

Portanguen, S., Tournayre, P., Sicard, J., Astruc, T., & Mirade, P. 2019. Toward the design of functional foods and biobased products by 3D printing: A review. *Trends in Food Science & Technology* 86: 188–198.

Rong, L., Chen, X., Shen, M., Yang, J., Qi, X., Li, Y., & Xie, J. 2023. The application of 3D printing technology on starch-based product: A review. *Trends in Food Science & Technology* 134: 149–161.

Rong, L., Shen, M., Wen, H., Xiao, W., Li, J., & Xie, J. 2022. Effects of xanthan, guar and Mesona chinensis Benth gums on the pasting, rheological, texture properties and microstructure of pea starch gels. *Food Hydrocolloids* 125: 107391.

Scaffaro, R., Citarrella, M. C., Gulino, E. F., & Morreale, M. 2022. Hedysarum coronarium-based green composites prepared by compression molding and fused deposition modeling. *Materials* 15 (2): 465.

Sher, D., & Tutó, X. 2015. Review of 3D food printing. *Temes de Disseny* 31 (31): 104–117.

Shahrubudin, N., T.C. Lee., & R. Ramlan. 2019. An overview on 3D printing technology: Technological, materials, and applications. *Procedia Manufacturing* 35: 1286–1296.

Shi, H., Li, J., Xu, E., Yang, H., Liu, D., & Yin, J. 2023. Microscale 3D printing of fish analogues using soy protein food ink. *Journal of Food Engineering* 347: 111436.

Sivamaruthi, B. S., Nallasamy, P. K., Suganthy, N., Kesika, P., & Chaiyasut, C. 2022. Pharmaceutical and biomedical applications of starch-based drug delivery system: A review. *Journal of Drug Delivery Science and Technology* 77: 103890.

Sun, J., Peng, Z., Zhou, W., Fuh, J. Y. H., Hong, G. S., & Chiu, A. 2015. A review on 3D printing for customized food fabrication. *Procedia Manufacturing* 1: 308–319.

Teng, X., Zhang, M., & Mujumdar, A. S. 2021. 4D printing: Recent advances and proposals in the food sector. *Trends in Food Science & Technology* 110: 349–363.

Tian, J. J., Bryksa, B. C., & Yada, R. Y. 2016. Feeding the world into the future-food and nutrition security: the role of food science and technology. *Frontiers in Life Science* 9 (3): 155–166.

Tiller, B., Reid, A., Zhu, B., Guerreiro, J., Domingo-Roca, R., Curt Jackson, J., & Windmill, J. F. C. 2019. Piezoelectric microphone via a digital light processing 3D printing process. *Materials & Design* 165:107593.

Truby, R. L., & Lewis, J. A. 2016. Printing soft matter in three dimensions. *Nature* 540 (7633): 371–378.

Tuan, D. Ngoa., & A. K. G. I. 2018. Additive manufacturing (3D printing): A review of materials, methods, applications and challenges. *Journal of Food Engineering* 143: 172–196.

Wan, X., Luo, L., Liu, Y., & Leng, J. 2020. Direct ink writing based 4D printing of materials and their applications. *Advanced Science* 7(16): 2001000.

Wang, L., Zhang, M., Bhandari, B., & Yang, C. 2018. Investigation on fish surimi gel as promising food material for 3D printing. *Journal of Food Engineering* 220: 101–108.

Wang, Y., Mao, B., Chu, S., Chen, S., Xing, H., Zhao, H., Wang, S., Wang, Y., Zhang, J., & Sun, B. 2023. Advanced manufacturing of high-speed steels: A critical review of the process design, microstructural evolution, and engineering performance. *Journal of Materials Research and Technology* 24: 8198–8240.

Wu, T., Chang, C., Chang, T., Chang, Y., Liew, Y., & Chau, C. 2019. Changes in physicochemical properties of corn starch upon modifications by atmospheric pressure plasma jet. *Food Chemistry* 283: 46–51.

Xu, D., Liu, Z., An, Z., Hu, L., Li, H., Mo, H., & Hati, S. 2023. Incorporation of probiotics into 3D printed Pickering emulsion gel stabilized by tea protein/xanthan gum. *Food Chemistry* 409: 135289.

Xu, K., Zhang, M., & Bhandari, B. 2020. Effect of novel ultrasonic-microwave combined pretreatment on the quality of 3D printed wheat starch-papaya system. *Food Biophysics* 15 (2): 249–260.

Yuk, H., & Zhao, X. 2018. A new 3D printing strategy by harnessing deformation, instability, and fracture of viscoelastic inks. *Advanced Materials* 30 (6): 1704028.

Zeng, X. X., Li, T. J., Zhu, J. C., Chen, L., & Zheng, B. 2021. Printability improvement of rice starch gel via catechin and procyanidin in hot extrusion 3D printing. *Food Hydrocolloids* 121:106997.

Zhang, H., Moon, S. K., & Ngo, T. H. 2020. 3D printed electronics of non-contact ink writing techniques: Status and promise. *International Journal of Precision Engineering and Manufacturing-Green Technology* 7 (2): 511–524.

Zhang, J., Li, Y., Cai, Y., Ahmad, I., Zhang, A., Ding, Y., Qiu, Y., Zhang, G., Tang, W., & Lyu, F. 2022. Hot extrusion 3D printing technologies based on starchy food: A review. *Carbohydrate Polymers* 294: 119763.

Zhang, L. L., Zheng, T. T., Wu, L. L., Han, Q., Chen, S. Y., Kong, Y., Li, G. C., Lei, M., Wu, H., Zhao, Y. H., Yu, Y. X., & Yang, Y. M. 2021. Fabrication and characterization of 3D-printed gellan gum/starch composite scaffold for schwann cells growth. *Nanotechnology Reviews* 10: 50–61.

Zhang, Z. J., Zheng, B., Tang, Y. K., & Chen, L. 2022. Starch concentration is an important factor for controlling its digestibility during hot-extrusion 3D printing. *Food Chemistry* 379: 132180.

Zheng, B., Qiu, Z. P., Xu, J. C., Zeng, X. X., Liu, K., & Chen, L. 2023. 3D printing-mediated microporous starch hydrogels for wound hemostasis. *Journal of Materials Chemistry B* 11 (35): 8411–8421.

Zheng, B., Tang, Y. K., Xie, F. W., & Chen, L. 2022. Effect of pre-printing gelatinization degree on the structure and digestibility of hot-extrusion 3D-printed starch. *Food Hydrocolloids* 124: 107210.

Zhuang, P., Greenberg, Z., & Mei, H. 2021. Biologically enhanced starch bio-ink for promoting 3D cell growth. *Advanced Materials Technologies* 6 (12): 2100551.

CHAPTER 11

Future Outlook

Enbo Xu, Dandan Li, and Yang Tao

As the second most abundant carbohydrate in nature, starch is the main component of the human diet, and also a potential source of functional additives such as thickener (Banerjee & Sarkar, 2015), resistant starch (Sun et al., 2022), carrier of bioactive compounds (Kimura-Todani et al., 2020) and commodity-like productions such as adhesive (Wang, Chang, Ma, Qiu, & Tian, 2021) and adsorbent (Guo, Wang, Zheng, & Jiang, 2019). It is a natural polysaccharide biopolymer obtained from renewable resources and is generally used for producing staple foods, snacks, bakery and fried products, batters, alcoholic beverages, sauces, soups, etc. Also, the use of starch for the production of non-food materials has been focused on its ability of being utilized as a matrix or additive (or functional materials) in many areas such as the environment, textiles, pharmaceuticals, fuels, adhesives, detergents, etc. Furthermore, the production of environmental-friendly starch-based products as a potential alternative to petroleum-based ones such as loose-fill packaging, film, sheet and foam for food and non-food uses have been considered for environmental protection.

This book describes the fundamentals of starch structure and their physical (sometimes with biological) modifications to develop edible, safe or functional materials. Mostly, novel starch-based (bio)materials are being developed with enhanced physicochemical, thermal, mechanical and optical properties. The inherent demerits of natural starch, such as low solubility, freeze-thaw stability, poor pressure, thermal and shear stability, and resistance to enzymatic hydrolysis limited its commercial application.

DOI: 10.1201/9781003493594-11

Starch modification began in the mid-16th century, during which period chemical modification is the most common processing method. However, chemical methods are facing intense scrutiny due to the negative impact of chemical agents on the environment, the additional cost of waste disposal and the toxic threat of human consumption. The growing demand for green/chemical-free labels by consumers and manufacturers has progressively driven the development of physical processing methods (Mhaske et al., 2022; Waseem et al., 2023). They have been successfully applied in many studies (hydrothermal treatment, ultrahigh pressure, extrusion, electric and magnetic fields, ultrasound, cold plasma, 3D printing and so on) to modify the composition, structure, function and digestibility of starch, which are summarized here (Chapters 2–10).

The function-oriented optimization and application have been focused on physical-treated starches as the main matrix and/or as an ingredient to enhance the final properties of food and non-food products in the last decades. The structural changes (from basic amylose/amylopectin chains to crystal/semi-crystal structure and even microscale granule morphology and surface) and utilizations of starches are achieved by different physical-based treatments. And more recently, the structure control of starch has been emphasized for realizing various functionalities, and among them, the advantages of nanosized starch with different morphologies (starch nano-sphere, nanorod, nano-worm, nano-polyhedron, nano-flake, nano-vesicle, nano-micelle and nano-networks etc.) are outstanding as for large specific surface area, good biocompatibility and non-toxicity. Many of them were efficiently prepared by physical methods which are green and sustainable, and have great potential in food, medicine and cosmetics applications. To summarize, physical treatment is a popular starch processing mode all along and will continue to maintain "vigorous vitality" of starch products.

REFERENCE, BIBLIOGRAPHY OR WORKS CITED

Banerjee, A., & Sarkar, A. K. 2015. Extraction, characterization, and evaluation of proso millet starch in textile printing. *AATCC Journal of Research* 2 (6) : 14–19.

BeMiller, J. N. 1997. Starch modification: Challenges and prospects. *Starch-Stärke* 49 (4): 127–131.

Davies, L. 1995. Starch-composition, modifications, applications and nutritional value in foodstuffs. *Food Technology Europe* 6 (7): 44–52.

Guo, J., Wang, J., Zheng, G., & Jiang, X. 2019. Optimization of the removal of reactive golden yellow SNE dye by cross-linked cationic starch and its adsorption properties. *Journal of Engineered Fibers and Fabrics* 14.

Kalambur, S., & Rizvi, S. S. H. 2006. An overview of starch-based plastic blends from reactive extrusion. *Journal of Plastic Film & Sheeting 22* (1): 39–58.

Kaur, B., Ariffin, F., Bhat, R., & Karim, A. A. 2012. Progress in starch modification in the last decade. *Food Hydrocolloids 26* (2): 398–404.

Kimura-Todani, T., Hata, T., Miyata, N., Takakura, S., Yoshihara, K., Zhang, X.-T., Asano, Y., Altaisaikhan, A., Tsukahara, T., & Sudo, N. 2020. Dietary delivery of acetate to the colon using acylated starches as a carrier exerts anxiolytic effects in mice. *Physiology & Behavior* 223: 113004.

Mhaske, P., Wang, Z., Farahnaky, A., Kasapis, S., & Majzoobi, M. 2022. Green and clean modification of cassava starch-effects on composition, structure, properties and digestibility. *Critical Reviews in Food Science and Nutrition 62* (28): 7801–7826.

Sun, H., Fan, J., Tian, Z., Ma, L., Meng, Y., Yang, Z., Zeng, X., Liu, X., Kang, L., & Nan, X. 2022. Effects of treatment methods on the formation of resistant starch in purple sweet potato. *Food Chemistry* 367: 130580.

Wang, F., Chang, R., Ma, R., Qiu, H., & Tian, Y. 2021. Eco-friendly and pH-responsive nano-starch-based superhydrophobic coatings for liquid-food residue reduction and freshness monitoring. *ACS Sustainable Chemistry & Engineering* 9 (30): 10142–10153.

Waseem, M., Tahir, A. U., & Majeed, Y. 2023. Printing the future of food: The physics perspective on 3D food printing. *Food Physics* 1: 100003.

Xie, F., Yu, L., Liu, H., & Chen, L. 2006. Starch modification using reactive extrusion. *Starch-Stärke* 58 (3–4): 131–139.

Xie, M., Duan, Y., Li, F., Wang, X., Cui, X., Bacha, U., Zhu, M., Xiao, Z., & Zhao, Z. 2017. Preparation and characterization of modified and functional starch (hexadecyl corboxymethyl starch) ether using reactive extrusion. *Starch-Stärke* 69 (5–6): 1600061.